THE BRAZILIAN AMAZON RAINFOREST

Global Ecopolitics, Development, and Democracy

Luiz C. Barbosa

University Press of America,® Inc.
Lanham • New York • Oxford

Copyright © 2000 by
University Press of America,® Inc.
4720 Boston Way
Lanham, Maryland 20706

12 Hid's Copse Rd.
Cumnor Hill, Oxford OX2 9JJ

Library of Congress Cataloging-in-Publication Data

ISBN 0-7618-1521-X (cloth: alk. ppr.)
ISBN 0-7618-1522-8 (ppr: alk. ppr.)

TO THE MEMORY OF MURIEL CLEMENT

Contents

Tables

Figures

ACKNOWLEDGMENTS

The writing of this book was made possible by a sabbatical leave granted by San Francisco State University in the Spring of 1998, and released time in the Spring of 1999. The university's School of Behavioral and Social Sciences also provided a small travel grant.

I express warm appreciation to Roland Clement, Clayton Dumont, Bill Howe, Rachel Kuhn-Hut, and Jonathan Purvis for their help in revising drafts of the manuscript. It was labor of love for each of them. Roland Clement twice edited the whole manuscript, greatly reducing my (Portuguese) prolixity. As with mistakes, what prolixity remains is my responsibility. My dedication honors Muriel Clement for introducing me to the possibilities of political protest while I was a student in her Connecticut home.

Chapter 1

Introduction

This book seeks to provide a global, world-systemic analysis of the problem of deforestation of the Brazilian Amazon rainforest. It shows how changes in global ecopolitics demanding sustainable development, coupled with the onset of democracy in Brazil, substantially altered the battle over the future of Amazônia. It views deforestation in the region in the context of an expanding frontier of global capitalism, and thus affected by capitalist interests. When global ecopolitics began to "green" in the 1980's and 1990's, Brazilian politics changed to accommodate the new realities. Destruction of the Amazon rainforest retains political and economic implications.

Beginning in the mid-1980's awareness of major global environmental problems, such as the greenhouse effect and the ozone hole, sparked an international debate on the state of the global environment; overnight the Brazilian government and those destroying Amazônia became environmental villains. The view of tropical rainforests as "green hells" quickly changed to that of "cradles of life" that must be preserved in the name of biodiversity and climatic stability. Suddenly Brazil was in the spotlight, pressured by environmentalists, international organizations, and first-world politicians to stop the vast devastation taking place in the Amazon rainforest, where between January 1978 and April 1988, 225,300 square kilometers (km^2) of forest had been cleared, an average of 21,130 km^2 per year (INPE, 1998, 1999). As tension mounted, the Brazilian government began to lose international financial allies in this process of destruction, e.g., the World Bank that had contributed millions of dollars in loans for environmentally unsound development projects. Under attack from environmentalists and politicians from major contributing countries such as the U.S., even these organizations began to pressure Brazil to avert the destruction, and threatened to cancel loans on controversial projects.

Meanwhile, people fighting for the preservation of Amazônia were gaining allies. The onset of democracy in Brazil politicized those few environmentalist and grassroots organizations already in existence, and facilitated the growth of hundreds of new ones.[1] These organizations in turn forged alliances with interested international non-governmental organizations (NGO's). The effort to preserve Amazônia quickly became globalized. Organizations such as the Rainforest Network and the Worldwide Fund embraced the cause of local grassroots groups, especially that of Native Brazilians, working with them to bring their plight and the devastation taking place in the forest to the attention of the world. Several grassroots leaders from the Amazon region, e.g., rubber-tapper leader Chico Mendes, and Indian leaders such as Davi Kopenawa, Chief Raoni, and Paulinho Payakan, were brought to countries such as the United States to influence loan decisions of politicians and other leaders. Public appearances in lectures, concerts, and television shows like the Phil Donahue Show and NBC's Today (see O'Connor, 1997: 218-222, 256) benefitted the parties involved: local grassroots groups gained political clout but, in turn, the testimonies of local leaders provided international environmentalist organizations with living examples of the consequences of development projects, thereby increasing their political leverage in, for example, U.S. congressional hearings.

By the late 1980's and early 1990's the effort to save the forests of the world was moving. The environment became the focus of major media attention, keeping the debate open and the heat on those on the side of destruction. *Time* devoted a whole issue to the state of the global environment, with much attention paid to tropical rainforests.[2] Several television documentaries were created, e.g., PBS Frontline produced "The Decade of Destruction," which examined the reasons behind Amazônia's devastation in great detail.[3] When some 400 Brazilian Indians gathered in the town of Altamira in the state of Pará to protest the building of the US$5.8-billion Kararao hydroelectric dam on the River Xingú, an equal number of environmentalists and journalists were also present (*The Economist*, March 11, 1989, p. 42). International celebrities joined the cause to "save the planet,"[4] thus associating their names with major environmental issues. The rock star Sting, for example, embraced the cause of the Kayapó Indians of Amazônia, and gathered US$1.2 million for the demarcation of their reservation.

Until a few years ago, rock stars who embraced a cause ran the risk of being considered lunatics . . . even in a world in which lunatics enjoy a certain prestige. Today this is not the case. Last week Sting, the pope of goodness rock, commandeered a show in Carnegie Hall in New York, with proceeds destined for the Rain Forest Network, an organization he created to denounce the torching of the Brazilian forests. Chief Raoni [of the Kayapó Nation], Sting's inseparable partner in his international crusade, was present and appeared on a big screen giving testimony--without translation. Among those present were Elton John, Caetano Veloso and Tom Jobim. The Brazilian Gilberto Gil participated also (*Veja*, March 20, 1991).[5]

Celebrity interest and the consequent coverage by the media gave new political leverage to both environmentalists and grassroots organizations, which now had a vehicle to deliver their message to an international audience.

More organizations devoted to the preservation of the forest and its native peoples emerged. According to the Documentation Department of the Indianist Missionary Council (CIMI), an agency linked to the National Conference of Bishops of Brazil (CNBB), by April 1993 there were at least 100 Indian organizations in Brazil alone, most of them created after the onset of democracy in the 1980's and located in Amazônia itself (CIMI, 1993a). Worldwide, there are over 219 organizations devoted to saving Amazônia (Trade and Environmental Data Base, n.d.). These organizations networked among themselves as to political strategy to influence decisions concerning the environment. Their use of the Worldwide Web influences an increasingly large public. When decisions viewed as environmentally unfriendly are made, they mobilize. They have been especially efficient in monitoring and putting pressure on the Brazilian government and international organizations regarding Amazônia. As we shall see, these organizations do not let the Brazilian government forget that there are negative consequences for environmentally unfriendly decisions.

As the debate on the state of the global environment intensified in the late 1980's and early 1990's, a new vision of development emerged: *sustainable development*, or development that is environmentally sustainable. Sustainable development became the catch word. Yet, in the beginning, its environmental emphasis was missing. An interview with economist Herman Daly illustrates his frustrations with most definitions:

I must say that academia has been no help. I compiled some definitions of sustainable development from the academic community that I thought were off-the-wall. Here are a few examples without attribution: "Sustainable development is development that sustains the highest rate of economic growth without fueling inflation." If that is any different than growth as usual, I can't figure it out. Another one says: "Sustainability is a concept taking into consideration the expanding needs of a growing world population, implying by this steady growth." That is just the same old thing. They didn't get it. They picked up the word, but not the concept behind it. Another talks about "a sustainable increase in rate of economic growth." Not just growth, it is acceleration that is going to be sustained in that definition. This is the kind of nonsense being bandied about in academia under the name of sustainability (in Lerner, 1991: 40).

The concept's usage has been influenced by ecological ideas such as "sustainability," "carrying capacity," "outer limits," and "maximum sustainable yield" (MSY) (see Munn, 1988:209). The term achieved an "official" character in 1987 in the report entitled *Our Common Future* by the World Commission on Environment and Development of the United Nations.[6] This fused the economic view of development with the eco-logical view of sustainability. For the Commission sustainable develop-ment meant ". . . a process of change in which the exploitation of resources, the direction of investments, the orientation of technological development . . . institutional change and the ability of the biosphere to absorb the effects of human activities are consistent with the future as well as present needs" (in Weston, 1995/1996). In many ways the concept is a compromise between radical preservationism that called for ecosystems like Amazônia to be left undisturbed, and radical development positions, especially by third-world leaders, claiming the right to use the environment for economic growth. It is a middle-of-the-road position that supports the use of ecosystems for development if certain guidelines are observed. Despite some vagueness it has been widely adopted, especially by international organizations such as the United Nations and the World Bank. It served to mute the argument for total preservation of vital ecosystems such as Amazônia. It legitimized the economic exploitation of ecosystems if done "properly" or "sustainably." It also dampened the debate over the sustainability of capitalism itself. The assumption implicit in the concept is that capitalism can somehow be tamed, that it can become environmentally sustainable if human beings adhere to certain environ-mentally sound practices. It elides the possibility that sustainability may not be achievable under the capitalist system and its free-market requirements. As the human population increases and capitalism becomes further globalized, demand for natural resources will increase (see Lyons et al., 1995:8-18). It will be tempting for corporations operating at a global level to take advantage of this situation. For example, as tropical forests have dwindled in southeast Asia, Malaysian timber companies have responded to global demands for timber by aggressively pursuing business in Amazônia (see *O Globo*, March 16, 1998). To what extent these companies will take advantage of the corruption endemic in Brazil remains to be seen. The Malaysian government has tried to assure the Brazilian Congress that Malaysian corporations will respect Brazilian law:

> The reaction against its country's corporations has mobilized the Malaysian Embassy. Last year, the then Ambassador Dato'Zainal Zain went to a session of the External Commission [Comissão Externa] in Congress for assurances that his countrymen intend to obey Brazilian law. According to him, the information about the devastation caused to tropical rainforests in his country was "biased and unilateral." It is part of an intimidation strategy by environmentalists to motivate international pressure against the corporations extracting the timber (*O Globo*, March 16, 1998, p.8).[7]

Where people are desperate enough for money or simply eager to enrich themselves, it is at least questionable to what extent environmental concerns will be able to tame the drives to survive and prosper.

Sustainable development achieved full recognition with the United Nations Conference on Environment and Development (UNCED), the "Earth Summit," held in Rio de Janeiro, Brazil, in June 1992. The Summit resulted from efforts of environmentalists, often against strong resistance by those who had much to gain from environmental destruction. It embodied a sense of realization that the current environmental problems afflict all countries, and that solutions require coordinated international efforts. Despite major disagreements between North and South, the Summit generated important treaties that officially changed, if only rhetorically, global ecopolitics.

Agenda 21, the main document signed in Rio, spells out how development is to take shape in the 21ST century. It specifies that ". . . environmental protection shall constitute an integral part of the development process and cannot be considered in isolation from it"(Principle 4, UNCED, 1992: 1). The protection of the environment, however, is for the sake of human beings who ". . . are the centre of concerns for sustainable development. They are entitled to a healthy and productive life in harmony with nature"(Principle 1, UNCED, 1992:1). Also, "the right to development must be fulfilled so as to equitably meet developmental and environmental needs of present and future generations" (Principle 3, UNCED, 1992: 1).

The conference was perhaps analogous to a ribbon-cutting ceremony;

it officially opened the gates to the new era of sustainable development. The rhetoric was in place. Despite the political fanfare of the conference, most of its promises remain unfulfilled. The "Rio Plus Five" conference held in New York in June 1997 showed that much needs to be done to implement the treaties signed in Rio. A major problem is that countries continue to disagree on major issues, particularly who will pay for the costs. Similar problems arose during the Conference of Climatic Change held in Kyoto, Japan in December 1997. There were major disagreements between the United States and Europe on timetables for carbon dioxide reduction. The main conflict centered around proposed reduction quotas for developing and developed countries. Developing countries like China argued that they should reduce less because of their development needs.

It is important to note the power of corporations in influencing the outcome of this decision. In the United States, for example, corporations bought a commercial on CNN arguing that all countries should reduce carbon emissions equally, otherwise the treaty would be unfair to the United States; that is, it would be unfair competition to U.S. corporations. The fact that the U.S. is the major emitter of carbon dioxide was omitted. In the end, it was decided that carbon dioxide emissions should be reduced to 1990 levels by 2008-2010. Countries would also be allowed to trade emission rights among themselves. Countries that developed carbon sinks, e.g., forests, would also receive credit (Wanick, 1997). These agreements should be viewed as a continuation of the UNCED treaties that provided the platform from which the Kyoto treaties could be launched. They reflect a growing awareness of global climatic change and the necessity of making the transition to more sustainable forms of development.

The debate about the state of the global environment meant that Brazil would become a special target of criticism and pressure. The preservation of the Amazon rainforest was regarded as vital due to its ecological importance and its size. The Amazon is believed to contain over half of the world's plant and animal species, still mostly uncatalogued. Two thirds of the seven million square kilometers of Amazon rainforest is within Brazil, making that country the biggest holder of the world's largest remaining tropical rainforest. Unfortunately, the rate of deforestation had been accelerating since the mid-1970's (see Table 1). By January 1978, 152,200 km^2 of Amazônia had been cleared and by April 1988 the figure had climbed to 377,500 km^2, or 225,300 km^2 in one decade. By 1989 another 23,900 km^2 was cleared, elevating the total cleared area to 401,400 km^2. The annual rate of destruction declined somewhat during the early 1990's, to 13,800 km^2 between 1989 and 1990, and to 11,200 km^2 between

Table 1. Extent of Deforestation in Amazonia (Km^2), 1978-1977

State	01/78	04/88	08/89	08/90	08/91	08/92	08/94	08/95	08/96	08/97
Acre	2,500	8,900	9,800	10,300	10,700	11,100	12,064	13,306	13,742	14,203
Amapá	200	800	1,000	1,300	1,700	1,736	1,736	1,782	1,782	1,846
Amazonas	1,700	19,700	21,700	22,200	23,200	23,999	24,739	26,629	27,434	28,140
Maranhão	63,900	90,800	92,300	93,400	94,100	95,235	95,979	97,761	99,338	99,789
Mato Grosso	20,000	71,500	79,600	83,600	86,500	91,174	103,614	112,150	119,141	125,023
Pará	56,400	131,500	139,300	144,200	148,800	151,787	160,355	169,007	176,138	181,225
Rondônia	4,200	30,000	31,800	33,500	34,600	36,865	42,055	46,152	48,648	50,529
Roraima	100	2,700	3,600	3,800	4,200	4,481	4,961	5,124	5,361	5,563
Tocantis	3,200	21,600	22,300	22,900	23,400	23,809	24,475	25,142	25,483	25,768
Legal Amazonia	152,200	377,500	401,404	415,200	426,400	440,186	469,978	497,055	517,069	532,086

Source: Instituto de Pesquisas Espaciais INPE (1999) <http://www.inpe.br/>

1990 and 1991. However, between 1991 and 1992 it rose again to 13,786 km^2. With the exception of 1994/1995, the overall rates remained lower than those of the 1970's. These rates, however, are unacceptable to ecologists, since they would eventually lead to the eradication of the forest.

The method used to clear the forest has been of special concern. Amazônia was being torched to the ground, to the point that, at times, forest fires could be seen with naked eyes by astronauts orbiting the earth. These fires have contributed to the greenhouse effect via carbon emissions. The clearing of tropical rainforests as a whole accounts for 15% of the effect of human greenhouse gas emissions (Stern, Young, and Druckman, 1992:67). The pictures of fires shown by the international media helped attract international public attention, especially as the reports linked them to global warming.

As criticism mounted against Brazil in the mid-1980's, and international organizations threatened to cut funds, the presidency of José Sarney (1985-1990) initially reacted with nationalistic belligerence, seeing criticism towards Brazil by the first-world countries as environmental imperialism. For President Sarney the developed countries could hardly criticize Brazil when they were the biggest contributors to the greenhouse effect through their carbon dioxide emissions. He was openly critical of their position, but his resistance could only go so far. Brazil is a country dependent on foreign investments, thus it had to do something to appease international public opinion. Program Nossa Natureza (Our Nature Program) was an effort by the Sarney administration to do just that—a nationalistic attempt to give the impression that Brazil was doing something to protect its forests (*Time*, September 18, 1989, p.85; Informativo FFCN, 1988). The clearest attempt to appease international public opinion, however, came with the administration of President Fernado Collor de Mello (1990-1992), who pressed to change the international image of Brazil as an environmental villain. He demarcated major Indian reservations such as the Yanomami and Kayapó in addition to taking other steps to reduce deforestation, such as enlisting army helicopters to monitor fires (see *Jornal do Brasil*, July 14, 1990). He also lobbied for Brazil to host UNCED. As we shall see, the environmentalist tone of the Collor administration was tied to Collor's economic agenda of integrating Brazil into the developed nations of the world. For him, Brazil should ally itself not with third-world countries but with the first world to which it aspired to belong.

The Collor administration specified its preferred method of preservation for Amazônia: the creation of Indian reservations, extractive reserves, and national parks. Roughly 396,000 km² or 8% of Brazilian forests are considered protected. In addition, Indian reservations encompass an additional 936,210 km² or 11% of the national territory; more than 200 were created after 1991 (Cabrera, 1996; Brazil 1997, Ch. 11; CEDI/PETI, 1990:6). The election of President Fernando Henrique Cardoso (1993-present) brought a few setbacks, however. On January 8, 1996 he signed Decree 1775, allowing the demarcation of Indian lands to be legally contested by corporations, farmers, and others. His administration apparently little realized the implications of such a decree. Non-governmental organizations were alert to the environmental decisions of the Brazilian government, and the decree created an uproar. They organized a letter-writing campaign on the Worldwide Web against the Brazilian government. They also put pressure on the governments of the G-7 countries and the World Bank so that they in turn would pressure Brazil (see InterPress Service, 1996; CIMI, 1996a; and Switkes, 1996).

While the changes in global ecopolitics described above were taking place, Brazil itself was going through major political transformations brought about by a gradual return to democratic institutions in the late 1970's, the period known as Abertura Democratica or Democratic Opening.

Democracy

It is our contention that within the context of an emerging global environmental culture, democracy is a key element for the preservation of the environment in the third world. As we hope to demonstrate with the Brazilian case, democracy can be a facilitator of preservation. It allows for battles between the capitalist forces of destruction and environmentalist forces of preservation to unfold. Authoritarian regimes such as military dictatorships do not allow the civil disobedience or dissent necessary for preservation groups to emerge and flourish. In these regimes the forces of destruction have the upper hand.

A major advantage of democracy for preservation is that it makes societies porous to the outside world, e.g., via free flow of information which allows diffusion of knowledge from the outside world and local events to become known. Local elites are forced to address problems they would otherwise ignore, certainly the case in Brazil in relation to Amazônia. Local populations are also connected to national and international debates via the media, and their perception can impact national politics through pressure on local politicians and support for local groups. Democracy thus allows the development of grassroots social movements. This has happened in Amazônia where Indians, rubber tappers, and landless peasants have organized as never before. The porousness created by the new freedoms has allowed these movements to form international alliances and bring their cause to the attention of the international public. Thus democracy has changed the nature of the struggle over Amazônia by allowing the forces of preservation to challenge the status quo. This situation differs markedly from that existing under the dictatorship (1964-1985), when civil disobedience was not allowed and dissidents were persecuted.

Another major contributions of democracy to the preservation of Amazônia has been the writing of a new constitution, promulgated on October 5, 1988. The *Constituintes*, or series of meetings and debates leading to the writing of this document, fostered a national debate on critical issues unique in Brazilian history. The process allowed for grassroots organizations to lobby for the passage of legislation sensitive to their needs. The interest of certain groups is reflected in the text of the Constitution itself, which has specific clauses devoted to equal rights for women, prosecution of racist acts, workers' rights, and so on.

Three chapters are of special importance for the future of Amazônia: Chapter III, "Agrarian Politics and Land Reform," Chapter VI, "The Environment," and Chapter VIII, "Indian [Rights]."

Chapter III calls for land reform by redistributing "public and fallow lands." Chapter VI states that Brazilians "have the right to an ecologically balanced environment . . ." and that it is required of government to preserve and restore essential ecosystems (Atr. 225, 1°, I). Chapter VIII recognizes Indians as having separate social organizations and cultures and affirms their rights to ancestral lands, calling for the demarcation of these lands (see Brazil, 1988).

Thus, the environmental chapter calls for a new environment-sensitive Brazil; the one on land reform calls for land to be given to the *"Sem Terras"* or landless peasants, who often end up in Amazônia where they continue to be landless; and, finally, the one on Indian rights calls for the protection of certain ecosystems for the benefit of those who have occupied them for thousands of years. These chapters are in the new constitution due to social movements, e.g., the Indian and the *"Sem Terra"* movements, which maintain pressure on the Brazilian government to implement them. These movements will seek to prevent the federal government from ignoring the constitutional mandates. Peasants often invade rural properties demanding land reform, thus calling attention to their cause through the national media, and Indians and their supporters have lobbied fiercely for demarcation and protection of their lands.[8]

Despite its importance as a facilitator of preservation, democracy alone cannot fully explain the movement toward preservation taken by Brazil, and by other countries. It needs to be examined in conjunction with global forces. What democracy did was to permit a dialectical process to unfold between divergent opinions and interests on the environment. This meant that the government, military or other, had to contend with forces in opposition to predatory capitalism.

The Dialectics

Ours is a dialectical approach to the problem of preserving the Amazon rainforest. Our approach is both Hegelian and Marxist; we borrow aspects of the dialects as proposed by both scholars. This approach sees the conflict between opposite forces as the primary mover of history, the impulse engine of social change. However, while Hegel's primary mover of history was ideational, Marx's was material:

> He [Hegel] seems to separate himself from Marxism by saying that what distinguishes man from beasts is the *"faculty of thought"*. Marx says that what separates man from beast is that man *labours*. But Marx would say that his thought is the result, consequence of his labour (James, 1980:27).

In the Hegelian approach new ideas lead to conflict and thus to change. For Marx, ideas also can lead to change, but they themselves arise from human beings' economic activities, they are a product of the means of production. New technologies and new relations of production or patterns of ownership create new classes with new aspirations and economic interests, the antithesis to the prevailing order, that clash with the status quo. The result is class conflict and its resolution moves history forward.[9]

Both Hegel's and Marx's dialectical views are useful in understanding the struggles taking place in Amazônia. While the primary force behind the struggles is economic, ideas also exercise a powerful force. For example, the gold contained in Indian reservations--such as the Yanomami-- explains why they are so often invaded. And, class conflict in other regions of Brazil, especially the Northeast, is responsible for the migration of thousands of peasants to Amazônia every year. However, to view the struggle over Amazônia strictly from an economic perspective is to have an incomplete picture of events, because ecological paradigms are challenging economic interests. Environmentalists both from abroad and within Brazil are using these new paradigms to question the traditional modernization views that have prevailed in that country. There is also the challenge posed by the ideas from the grassroots social movements of the people of the forest themselves. Indians, for example, have argued that they are the "true protectors of the forest," that their way of life is better ecologically than that of "Brazilians." This view originally came from the outside world, perhaps largely constructed by the global media. Also, sustainable development is itself an "idea" that has the potential of changing the way we live. It requires economic rethinking and restructuring. It assumes that certain ecosystems should be maintained or managed for their ecological rather than economic value in order to benefit all people currently living as well as those of future generations. Above all, sustainable development provides us with a sense of ecological rationality to balance against the irrationality of capitalism itself: the irrationality of sacrificing vital ecosystems such as Amazônia for the sake of consumerism.

In addition to borrowing from both Hegel and Marx, this analysis uses a global dialectical approach; it assumes that changes have taken place in the ecopolitics of Amazônia as a result of conflict between opposing forces both within Brazil and abroad. We argue that there is a battle taking place over Amazônia between capitalist (systemic) forces of destruction, and the anti-systemic forces of preservation and conservation.[10] Systemic and anti-systemic forces are both domestic and international. It is a battle between, on one side, the Brazilian government, the Brazilian landed elites, domestic and international corporations and international lending institutions and; on the other side, Indians, rubber tappers, grassroots organizations and NGO's. Even though the specific battle is new, the conflict itself has taken various forms since the discovery of Brazil by the Portuguese in 1500. Indians, for example, have resisted the penetration of Europeans into their lands from the beginning. Globalization and democracy changed the dialectics over Amazônia. The anti-systemic side increased in strength, but the systemic forces still retain the upper hand. It is, however, no longer business as usual.

Thus, for us, what is taking place in Amazônia is part of something bigger, part of a global dialectical process over the state of the environment. Despite its limitations, *sustainable development* does present a challenge to the prevailing paradigm of predatory capitalism. Specifically, it challenges the capitalist consumerist culture that is spreading throughout the world (see Durning, 1992). Changes in the ecopolitics over Amazônia are a reflection of these overall changes in global ecopolitics. The support of ecopolitics for sustainable development have sent shock waves throughout the world system. The effects on the ecopolitics of third-world countries have been particularly dramatic, especially for those with democratic governments. These changes, "greening steps" toward a new economic paradigm of sustainable development, rather than radical revolutionary political or economic changes, may well be the mode of transition to the new age. The frequency and severity of environmental problems will to a large extent determine the "greening steps" we shall take in the future.

The Dialectic and NGO's

Central to the dialectical process on a global scale is the presence of an increasing number of non-governmental organizations (NGO's) devoted to saving the planet. From 1909 to 1988 intergovernmental organizations grew from thirty-seven to 309, while NGO's grew from 309 to 4,4518. The number of NGO's devoted to saving the environment varies from directory to directory. The Environmental Liaison Center International (ELCI) connected with the United Nations Environmental program (UNEP) listed 726 organizations in 1993. On the other hand, The International Directory of Non-Governmental Organizations listed some 1,650 environmental and development NGO's in its 1992 edition. Whatever their number these organizations have become major players in the international politics of development. Their participation in major international conferences such as UNCED (Princen and Finger, 1995:2-4) shows this. While their role in international politics is still evolving, it is evident that they have contributed substantially to democratizing development worldwide. They argue against the prevailing style of development promoted by international organizations that has been insensitive to the needs of ordinary people and the environment. They have promoted the aspirations of common people, thus opening discussions to more democracy (Clark, 1991:4). In *Democratizing Development*, Clark writes:

> Indeed, the focus on wealth production rather than wealth distribution, production for export rather than for the needs of local populations, extraction of natural resources rather than environmental protection and Western-style technologies, for example in agriculture, have often compounded the problems we now regard as critical. . . . There is now near-universal recognition that poverty alleviation, eradication of hunger, protecting the environment, grassroots development and safeguarding the poor from the debt crisis are priorities. But these are areas in which the

official agencies have limited experience or discover tremendous operational difficulties and as a result they are actively seeking the collaboration of both Northern and Southern NGOs. The voluntary organizations often work in the areas of greatest poverty, have direct relationships with the communities of the poor people and have considerable experience of tackling environmental problems (1990:4-5).

These organizations and their networks have created a substantial challenge to the status quo. On a global level they monitor and challenge the decisions made by official organizations such as the World Bank. By opposition to traditional views of development, by pointing out their consequences, and by representing the interests of the poor, they have questioned the very rationality of major decisions.

The Dialectic and Rationality

Our dialectical view implies a struggle between the material and the non-material, between economic interests and ecological views. From this perspective the struggle over ecosystems such as Amazônia involves a struggle against the irrationalities of capitalism. Capitalism itself is dialectical, it is both rational and irrational. On the rational side:

> Capitalism has hitherto involved the development and application of science and technology by hierarchical, bureaucratic organizations whose relationships have been governed predominantly by the principles of exchange on the market. It has consisted of rationality in the sense of planning by private corporations and public state agencies in terms of profit and loss, supply and demand, resource extraction and consumption (Murphy, 1994:37).

However, on the irrational side:

> Unforeseen side-effects have been encountered. New unknowns have appeared in the calculations. In the cumulative process of the discovery of scientific, technological, and organizational solutions to problems, further problems of increasing severity have accumulated as well: radioactive and toxic waste, pollution of the oceans and the atmosphere, high technological accidents, increasingly destructive weapons systems, destruction of forests and depletion of resources, extinction of species, etc. (Murphy, 1994:.38).

What appears to be rational solutions, e.g., development of nuclear energy, can have irrational outcomes such as radioactive waste. What is rational and irrational often depends on the stance one takes. For making decisions on industrialization, nuclear plants can seem the rational course of action. From the perspective of potential loss of human health it is utter irrationality. From the perspective of profits and the ideology of development it makes sense to open tropical rainforests. From a broader ecological view of environmental degradation it does not; it is irrational to destroy such ecosystems when humanity is faced with problems like the greenhouse effect and the hole in the ozone layer.

Amazonian Development and Irrationality

As we shall see throughout this book, the models of development employed in Amazônia have had dangerous ecological consequences. These models have been devised by Brazilian technocrats as a rational way of incorporating Amazônia into Brazil's capitalist economy. In these models, since the land was cleared of jungle and put into productive use, deforestation was viewed as "improvement," as development. The un-sustainability of most economic activities selected for the region was ignored under the premise that Amazônia is so vast that it could never disappear. For example, despite ample evidence that cattle ranching is not a sustainable economic activity for most of the region, it received both federal economic incentives, and investments from abroad (Campuzano, 1979). Thanks to these subsidies, wealthy investors benefitted in the short run while huge portions of the vital ecosystem have been destroyed for the future. In addition to cattle ranching, other non-sustainable activities in the region, such as mining, logging, and even the settlement of people, have been carried out. Gold prospectors use mercury which severely contaminates local rivers.

Brazil has resisted criticisms of its models of development for Amazônia. The rationale is national sovereignty and security. Amazônia belongs to Brazil and Brazilians can do whatever they wish within their own country. Development is a "benefit" for developing countries. A statement by former Brazilian Ambassador to the United Nations, Paulo Nogueira-Batista, illustrates this point of view:

> The right of all nations to develop came to be recognized by the international community as a natural consequence of the political decolonization process which took place after the II World War. The independent under-developed countries were given to understand that, to exercise that fundamental right they were entitled to resort to the same technological solutions used firstly by the market oriented economies of the West and subsequently by the central-planned economies of the East (1989, p.1).

While the claim for equal rights for the Third World is legitimate, it ignores the different historical circumstances surrounding the development of the West. Being the first to undergo the Industrial Revolution, the West had plenty of "polluting space" (Barbosa, 1990) and resources to use. Today, given the green house effect and other major global environmental problems, this space is no longer available and forests are needed for climatic stability. In addition, this way of thinking ignores the fact that the Western model of development created the ecological disasters we now have. Given the development of science which accompanied western capitalism, we now have understandings not available during the industrial age of the West. It makes little rational sense to pursue a model of development that will generate the same ecological harmful effects it generated elsewhere. The West's predatory capitalism did away with most of its forests, a process still unfolding in northwest United States and Canada. Brazil could be at the forefront of sustainable development by using a more adequate model for Amazônia. It is our position that development in the region should be kept to a minimum due to the fragility of its ecosystems. Perhaps it should remain limited to extractive activities such as rubber tapping. It is also our position, however, that the same should apply to other remaining forests in the world, including those in Alaska and Canada. Considering the global environmental problems of our time it makes no sense to continue to plunder them.

GLOBALIZATION: A WORLD-SYSTEMIC VIEW

In an analysis guided by a world-systemic approach, as here, globalization has a definite meaning. World-system theory is built on the premise that the current capitalist world-economy has its roots in the European expansion of the "long-sixteenth century" (1450-1650). According to Immanuel Wallerstein, a leading world systems scholar, it was in this period that the capitalist European economy spread to engulf all corners of the world. This world-economy differed from "world empires" because it was based on economic exchange rather than political domination. Its expansion joined people of different cultures under the same economic system, but these societies were incorporated into the system differently. As capitalism spread, through colonialism, a global division of labor emerged in which the western European countries became the center or core of the economy, and the poor colonial societies its periphery. Between the core and the periphery a third category, or semi-periphery, emerged consisting of declining core countries and a few, such as the United States in the 19th century, which managed to escape peripheralization. In the twentieth century the United States, and later Japan, joined the core. The relationship between the core and the periphery has been one of exploitation of natural resources, labor, markets, etc. For example, as the New World was conquered and demand for goods increased, virgin forests were opened for agriculture, Indians were enslaved, African slaves were exchanged, silver and gold mines were opened, etc. That is, demand in the core regulated the economy of the periphery, shaping its socio-political landscape. Today, peripheral countries provide multinational corporations with an army of cheap labor. It is no coincidence that so many multinational corporations have assembly plants in places such as Mexico and Southeast Asia where there are people "willing" to work for meager wages. The relationship with the core has kept the peripheral countries poor because resources needed to build internal infrastructure have been extracted from them. Thus, world-system scholars argue that globalization is not a new phenomenon. It has been unfolding since the "long sixteenth century." Today's "globalization" is simply another stage in the evolution of an evolving capitalist world-economy.

Globalization has become a catchy word to describe the compression of the world, the fact that we are living in an increasingly interconnected world. Despite the agreement among scholars as to an overall trend towards globalization there is much disagreement on the extent and nature of the process itself. A principal area of disagreement is what constitutes the moving force behind the trend. Is globalization brought about by the spread of international trade, as proposed by world-system scholars? Or do politics and culture play a primary role in the process of globalization as well? A trend in opinion seems to be moving from traditional economic explanations of globalization to include culture and politics as independent of economic processes, and more often than not governing them. This trend has been spearheaded by sociologist Roland Robertson, "the key figure in the formation and specification of the concept of globalization ..." (Waters, 1995: 39). For him, the most important trend in globalization today is an increasing global consciousness among people and the development of genuine global culture:

> The global field as a whole is a sociocultural 'system' which has resulted from the compression of--to the point that it increasingly imposes constraints upon, but also differently empowers--civilizational cultures, national societies, intra- and cross-national movements and organizations, sub-societies and ethnic groups, intra-societal quasi-groups, individuals, and so

on. As the general process of globalization proceeds there is a concomitant constraint upon such entities to 'identify' themselves in relation to the global human circumstance (Robertson, 1992: 61).

Robertson's approach is critical of world-system theory due to that theory's economic emphasis:

> In any case, even if we were to agree about the prime-mover significance of 'the economy' in the making of the modern world, that does not in and of itself lead to the simple conclusion that - to take up the major problem - 'culture' has been epiphenomenal. Wallerstein's record with respect to the central issue is not entirely clear-cut. For long, he explicitly stated that 'culture' is to be regarded as epiphenomenal. Yet he has devoted significant portions of his major works to the discussion of cultural, particularly religious, matters . . . (Robertson, 1992: 65).

For many scholars, therefore, culture and politics constitute fundamental components of globalization analysis. They argue that innovations now spread quickly throughout the world via such means as television, the internet and other forms of communication. Government leaders also pay attention to international organizations, other states, and international public opinion or risk the consequences, such as embargoes, boycotts, high tariffs, loss of preferential trading status, etc. Often, politics contradict economic rationale, as in the case of the Iranian Revolution of 1979 which severed Iran's major economic links with the West.

The emerging emphasis on culture and politics is indeed a central component of globalization. One cannot ignore the importance of cultural diffusion in changing the behavior of people everywhere toward consumerism. However, for us, economic forces remain a "prime-mover" behind globalization. Globalization means above all the westernization of the world via the spread of capitalism and its consumerist culture. A critical feature of the current global system is that not all countries benefit equally economically, nor do they make an equal contribution to global politics and culture. Global politics are shaped by the economic interests of the rich countries and their multinational corporations. The same applies to global culture. This culture has spread via the diffusion of a consumerist mentality, so necessary for the survival of capitalism. National medias are dominated by the western media, especially by U.S. media corporations. In the third world, television is heavily loaded with western programs, wherein people have access only to a glamorized version of western (mostly American) culture. This involves a standard of living the vast majority of third-world people can only dream about. This does not mean that all global processes can be reduced to economics. There are still major cultural, religious, and political differences which affect the dynamics of the system. However, these are constantly being bombarded by capitalist interests, which attempt to transform and align them along acceptably capitalist modes. For example, "traditional" cultures are constantly being infiltrated and westernized. Thus, one cannot fully understand globalization without recognizing the nature of capitalism itself; the infrastructure of the globalization process is the capitalist economy.

Capitalism is driven by the profit motive, it entails the commodification of everything, including human beings in the form of labor. Things only have economic value in the system if they can generate profits. Nature is viewed as a source of raw materials which acquire value only when transformed into commodities to be sold in the market.[11] Preserving nature for the sake of nature is a concept often antithetical to the ideals of capitalism. Such a principle would prevent capitalists from using eco-systems as sources of raw materials--perhaps an exception is the burgeoning field of eco-tourism. The expansion of capitalism meant that its assumptions would spread to different corners of the world, by force or voluntary adoption. Thus, when Europeans arrived in the New World they brought with them the concept of land as private property, a concept different from that of Native Americans who saw land as communal. Because Native Americans resisted the takeover and the commodification of their lands they became an obstacle to the profit motive and had to be removed where they could not be assimilated into the system. This process continues today. The Brazilian government's official policy towards Native Brazilians has been one of "civilizing," "assimilating" or "integrating" them to the rest of the country. Their lifestyles are perceived as incongruent with the modern developmental needs of the nation.

Commodification is central to the ideology of progress in capitalism: things must have monetary value in order to be marketable (Wallerstein, 1990, Ch. 2; 1987:15-16). Thus, land is valued only when it has been partioned and become private property that can be sold and resold; that is, when it becomes a form of wealth. Progress is defined to a large extent in terms of the production and consumption and technological advancements. The assumption is that science, so far a major ally of capitalism, is capable of producing more advanced technologies to further exploit nature and always make the future better than the present. Preservation of nature is often perceived as an attempt to maintain nature "idle," so to satisfy the curiosity of a small group of environmentalists at the expense of progress and the common good. Native peoples and their cultures are viewed as obstacles to progress, especially if their lands contain natural resources that their cultures do not value in the same fashion as do "modern," capitalist cultures.

The industrial revolutions of Europe and the United States accelerated the process of commodification of nature. Goods were produced en masse for an increasingly larger market. Nature was converted into material goods as never before in human history. The Industrial Revolution also magnified the faith in progress which we inherited from the Enlightenment (see Collins, 1994, Prologue). Human progress was no longer in doubt in philosophical discussion; progress, as defined in capitalist terms, was unfolding in industrialization, urbanization, modernization, etc. As other parts of the world were incorporated into this European economy they were transformed. Colonialism changed the ways people in the periphery saw themselves, and the future to which they aspired for themselves and their children. It placed Europeans at the top and the natives at the bottom of an economic, moral, and ethnic pyramid. Peripheral people themselves came to believe that Europeans and their culture were superior, and that "traditional" ways were therefore inferior or backward. While Europeans had access to wealth, prestige, and power, the locals worked in manual jobs often catering to the needs of colonial masters. A better life became associated with being westernized, with adopting western culture and values. When colonies obtained independence they more often than not looked to the West for their model of economic development. Industrialization and the creation of a consumer society became the ultimate goal for these countries, a goal still chimerical for most of them.

Political independence for the former colonies did not translate into economic independence, since they continued to be peripheries in the capitalist world-economy. Colonialism transformed itself into neo-colonialism; that is, political "independence" with continued economic exploitation, since peripheral countries depend on the markets of the core for survival. The majority continue to participate in global capitalism as producers of raw materials and agricultural goods. Some countries attempted to escape this condition by a concerted effort to industrialize.

In Latin America this effort came under the banner of import substitution. In import substitution government assumed the role of fomenting economic development that the bourgeoisie assumed in the economic development of the West. The aim was to create and protect local industries whose goods would "substitute" for those that were imported. It also involved the creation of infrastructure such as roads, railroads, and hydroelectric dams to support local industry. For several countries this attempt at industrialization led to what Peter Evans (1979) has called dependent development, development dependent on foreign capital. Even Brazil, which achieved significant levels of industrialization by adopting this model, had to borrow heavily from international institutions. The third world's attempt to catch up with the West incurred a foreign debt in the trillions of U.S. dollars.

Brazil's achievements continue to be dependent on the core because it relied heavily on international capital to finance development projects in places such as Amazônia. By 1997, it owed more than $120 billion dollars, making it the biggest third-world debtor nation. This debt has exacerbated many of the country's social ills of poverty, unemployment, child abandonment, and crime. It has had environmental implications as well. Brazil has sold its natural resources cheaply in international markets in order to obtain hard currency to service its debt (see Danaher and Shellberger, 1995).

Foreign debt has meant increased poverty for many parts of the world, to the point that scholars have called the 1980's the "lost decade" of economic development (see Raymond, 1991). The money that could have been invested in local infrastructure, e.g., schools and hospitals, has been sent to the core instead (see Onimode, 1989). Poverty causes environmental degradation since the poor often destroy forests in search of land, cooking fuel, food, and so on.

Global Capitalist Institutions

It is important to note that neo-colonialism involved the creation of international institutions to promote and finance the West's model of development. A good example is the World Bank. While originally created to address the redevelopment of Europe in the aftermath of WWII, in the 1950's and 1960's the Bank assumed the role of financier of development projects in the Third World. Other institutions such as the International Monetary Fund (IMF) were created to stimulate capitalist development. These institutions naturally represent the interest of the rich countries which finance them and thus have majority voting power. They have helped disseminate the ideological views of progress inherited in the capitalist system. For them, tropical rainforests are idle regions containing natural resources that can be used to pay for the loans contracted by developing countries. Their financing of environmentally destructive projects in the name of capitalist development makes them systemic agents of destruction (Barbosa 1993a; and 1993b).

To world-systemic organizations, the adoption of capitalism was essential if a country was to receive economic assistance. For them, development meant modernization along capitalist lines. Both the World Bank and the IMF have operated along these lines. The Bank, for example, has been criticized on many occasions for its emphasis on large infrastructure projects which have benefitted corporations (and the rich) the most. From the Bank's perspective the poor would eventually benefit from such projects by having access to the jobs created by corporations, in addition to using the roads, electricity, etc.; that is, the benefits would "trickle down" to them. Critics argue that, instead, these projects have had devastating consequences for the poor. They have entailed the loss of land, relocations, and pollution of their environments (Helmore, 1988). The policies of the IMF have had similar adverse effects. They have called for economic stabilization, especially reduction in inflation rates via lessened government spending on social services, privatization, exports, free markets, etc. These strategies often come in the form of "conditionalities" imposed on developing countries who request loans, or the seal of approval by the IMF to obtain loans from private lenders (Onimode, 1989). The result has been that austerity programs have been shouldered most heavily by the poor. They are the ones who lose food subsidies, health care, and new schools for their children. Also, large landowners attracted by the lure of larger profits switch from production for the domestic market, e.g., food, to the production of goods to be sold abroad, e.g., coffee. Peasants may lose their livelihood when, in order to be competitive in the international markets, mechanization of production is introduced. Programs of stabilization require repayment of debt to foreign lenders. In order to obtain the hard currency needed for this, poor countries sell their natural resources at banana prices. As we shall see in Chapter III, this has certainly been the case of the Carajás Project in northeastern Amazônia, which was at first viewed as "salvation" for Brazil's mounting foreign debt.

Thus, from a neo-Marxist point of view, world-system theory sees the world as a system of inequality, one in which the rich countries have control through the power of money held by their transnational corporations, international organizations, and banks (McGrew, 1992: 21-22). Global politics in this approach are viewed as being shaped by the interests of core countries, especially by the dominant or hegemonic state of the time. Economically, the periphery is exploited by the core and politically its situation is one of subordination. Nationalism is subverted. Resistance occurs but because they control much of the world economy the core countries retain the upper hand. Countries dependent on foreign capital, such as Brazil, eventually succumb to external pressure. In addition, their history of attachment to a capitalist world-economy explains much of their "underdevelopment." Colonialism meant that their resources flowed to the core, e.g., the silver of Mexico to Spain. Neo-colonialism continues this exodus of capital via dependent development and interests paid on external debts. For Wallerstein (1995) this condition of inequality will always exist in the system:

> Thus, it is true on the one hand that some so-called national development is always possible, indeed it is a recurrent process of the system. But it is equally true that, since the overall maldistribution of reward is a constant, both historically and theoretically, any "development" in one part of the world-economy is in fact the obverse face of some "decline" or "de-development" or "underdevelopment" somewhere else in the world-economy. And this was less true in 1893 than it was in 1993; indeed it was no less true in 1593. So I am not saying that it would be impossible for x-country to "develop" (today, yesterday, or tomorrow). What I am saying is that there is no way, within the framework of our current system, for all (or even many) countries simultaneously to "develop" (p. 167).

The world-system, in this view, is a rat race. But in this race everyone tries to be like core countries in their consumption lifestyle and political power. Unaware of "their place" in the world-economy, in the process of trying they dilapidate their environment by polluting it, clear-cutting its forests, removing its minerals, damming its rivers, etc. And, this has been done with the help of international organizations such as the World Bank which finance the destruction. Not that these organizations are solely responsible for these problems. However, their assistance and disregard for environmental consequences have magnified and accelerated them.

Unincorporated Areas and Resistance

Despite the European world-economy's steady engulfment of other parts of the world through colonialism, several regions were either loosely incorporated into the system or left unincorporated altogether until recently. Thomas D. Hall (1986), for example, has shown how the Indigenous peoples of North America were integrated differently into the system. He provides us with a continuum of four categories of incorporation: None, Weak, Moderate, and Strong (p. 392). And, according to him:

... incorporation is a matter of degree and can be a volatile process. Shifts in incorporation are not entirely elastic. As a region becomes more closely articulated to the world-economy, external pressures impinge more forcefully on local groups. When such pressures are sufficiently strong, and of sufficient duration, the structure of local groups is changed. If the transformation is drastic it becomes more difficult to reverse--even in the event of a shift to looser incorporation (1986:398).

The introduction of the horse was one of those irreversible cultural adaptations to incorporation. For him, uncontacted areas, non-incorporated, are "external arenas" to the capitalist world-economy. Areas with weak links are "contact peripheries," those with moderate links are "marginal peripheries or regions of refuge," and those with strong links are called "full-blown" or "dependent" peripheries (Hall, 1986:392).

Despite its North American emphasis, Hall's model provides a good tool for understanding the situation taking place in Amazônia. As a frontier region it is a region of contact and growing incorporation. In Amazônia one finds all of Hall's categories. Small towns and cities are good examples of full-blown peripheries, and Indian reservations and villages in development areas good examples of contact peripheries. There are also areas in the region which lie totally outside the world system, with no links to the capitalist world-economy. Their people live in almost complete isolation:

Brazil is one of the few countries in the world where there are still ethnic groups which have not had contact with society at large. They are what we call isolated Indians. According to a FUNAI survey, all of them are located in the Amazon, with the exception of the Ava-Canoeiro in Goiás (Brazilian Embassy, 1993:7).

In this book, we view incorporation as resulting from three main processes: 1) Nature of Contact; 2) Degree of Resistance; and 3) Military Technology. The Nature of Contact ranged from peaceful to warlike, often oscillating between the two. Both had devastating consequences for the natives. Due to lack of immunity to disease both war and peaceful contact killed the natives. Historically, some groups have resisted contact more than others. When faced with the presence of Europeans some chose to penetrate farther into the forest. Others chose to resist militarily, by attacking the newcomers with bows and arrows. The lack of efficacy of the resistance was determined by the superior technology of the newcomers and the lure of the goods they brought with them. Bows and arrows are no match for guns, and it has been difficult for some Indians to resist the goods used to lure them into contact. In Brazil, for example, "hooking" Indians with modern goods has been a way to assimilate them into the fringes of Brazilian society. The avoidance of contact in the latter part of the twentieth century has become a costly option for them:

> A few Indigenous groups managed to reach the present day maintaining their autonomy despite the pressure on their territory, and they received special protection from FUNAI, through its Isolated Indian Department. The most recent survey of those groups identifies 75 isolated groups. All of them, in one form or another, have begun to feel the impact of encroachment--especially from prospectors, lumberjacks and settlers. The expansion of national society has led those groups to try to find strategies to guarantee their survival, from holding their ground and fighting to looking for new areas where they can maintain their autonomy (Brazilian Embassy, 1993:7).

This book describes the expanding frontier of capitalism in Amazônia, an expansion which has taken place in sporadic waves throughout history. Since the 1960's, however, the forest has been hit by enormous and constant waves of colonization and "development." These tsunamis of incorporation now threaten the very survival of the forest and its traditional inhabitants. Incorporation has meant displacement, or the threat of displacement, and loss of land for the people of the forest. They have resisted as best as they could. In the process of resisting they have been transformed. Since this dilemma was created by the expanding frontier of global capitalist, solutions will require global changes.

Notes

[1.] According to Eduardo Viola (1988) there were a few environmentalist organizations in Brazil during the period of military dictatorship. These organizations to a large extent devoted themselves to the task of environmental education, abstaining as much as possible from politics. The reason for this is obvious. Criticism of the military system entailed political repression. Being apolitical was a strategy of survival.

[2.] The end-of-the-year (1988) issue of *Time* was entitled "Planet of the Year," instead of its regular "person-of-the-year" issue (*Time*, January 2, 1989). For a discussion of this topic see also Kraft and Vig (1990).

[3.] This documentary was produced by Adrian Cowell and is one of the most comprehensive documentary analyses of the region. It examines the problems in Amazônia from different perspectives, e.g., it has a whole episode, "In the Ashes of the Forest," on land disputes and armed conflicts plaguing the region.

[4.] Actress Maryl Streep embraced the cause by narrating a whole documentary, "Race to Save the Planet," which addressed different aspects of the global environmental crisis.

[5.] Author's translation.

[6.] For a history of the development of the term see Redclift (1987). For the use of the term by the World Commission on Environment and Development see Weston (1995/1996).

[7.] Author's translation.

[8.] It has also become a common strategy in the 1990's to invade government offices. This is a strategy frequently used by the landless but it is also used by Indians, who have invaded the Fundação Nacional do Índio (FUNAI), the Brazilian equivalent to the B.I.A., offices. For example, a group of 170 Indians invaded a FUNAI building in Campo Grande, Mato Grosso do Sul on April 7, 1998 to demand that a non-Indian be replaced by an Indian as the local FUNAI administrator (*Jornal do Brasil on Line*, April 8, 1998). A very bold move by the landless was the threat to stop the stock market of São Paulo, the financial heart of the country (see *Jornal do Brasil*, March 19, 1998).

[9.] See the excerpts from Marx's *The German Ideology* (pp.23-28) and *Capital III* (pp. 217-221) in John Ester (1986).

[10.] For us there is a clear distinction between preservation and conservation. Preservation means to preserve as is while conservation means to use resources wisely, e.g., by replanting forests.

[11.] Even Karl Marx himself who was so critical of the capitalist system took nature for granted. For him, raw materials were given to us "gratis" by nature and acquired value as they were transformed by human labor. In line with the thinking of his time, Marx did not acknowledge the possibility that nature could have value on its own right, as a source of life (Marx, 1906).

Chapter 2

Historical Links: Amazônia Before the Mid-1980's

Cycles and Sub-Cycles

According to Brazilian economist Mircea Buescu (1970) Brazilian socio-economic history can be understood as a series of economic cycles and sub-cycles. Cycles lasted longer, centuries or near half centuries, and sub-cycles lasted for shorter periods. The latter also tended to be regional in scope, albeit having a substantial impact on the revenues of the colony and, after independence on September 7, 1821, of the country. Cycles and sub-cycles shaped the very fabric of social life in Brazil. People organized their social lives around these economic activities. African slavery, for example, became a major feature of Brazilian social life as a result of the plantation economy that characterized Brazilian agriculture. The major economic cycles were Brazilwood, 1503-1550, Sugar, from 1550 to 1650, Mining of Gold and Precious Stones, from 1694 to 1760, and Coffee, 1825 to 1930. The major sub-cycles were Cattle Ranching from 1560 to the end of the colonial era, Tobacco from 1642 until the end of the colonial period, Cotton 1780 to 1790, and Rubber from 1880 to 1912 (Buescu, 1970:42, 243).

This cyclical economy was triggered by markets in the world economy, especially European demand for tropical goods. Thus, large tracts of Atlantic coastal rainforest were cleared to provide brazilwood for the production of textile dyes, to the point that brazilwood became a rare species where it once thrived (see Dean, 1983). The same applies to the sugar and coffee cycles, when forests were cleared for the creation of plantations for the production of these commodities. Brazilian colonial and post-colonial history until 1888, when slavery was abolished, can be viewed as the history of expansion and contraction of an economic (mostly slave-holding) system directly linked to Europe. When there was a demand for Brazilian products in Europe new lands were put under cultivation, African slaves were imported, Indians were enslaved (see Lang, 1982). When the demand declined the opposite happened: plantations stagnated and the manumission of slaves became common (see Degler, 1971).

Relative to the rest of Brazil, the Amazon region was an economic backwater, an extreme periphery. Until the mid-1700's the region was divided into several missions under the strict control of the Jesuits, who enslaved the Indians under the banner of civilization and Christianity. In addition to the production of food staples for local consumption using Indian labor, the main economic activity was the extraction of the *drogas do sertão* (drugs of the interior), mainly spices such as clover and cinnamon. It was not until the mid-1700's, under the more capitalistic administration of the Marquis de Pombal, that attempts were made to develop export agriculture in the region. Rice, cacao, coffee, and cotton produced in large plantations using slave labor became major commodities exported from the region, primarily from what is now known as the states of Pará and Maranhão (Hemming, 1987: 42-43). However, much of Amazônia remained untouched by these economic activities which tended to be concentrated near large urban centers such as Manaus and Belém. Nevertheless, even such restricted economic activities had dire conse-quences for the Indians. When the Portuguese arrived in Brazil there were an estimated 2.5 million or more Indians. By the mid-eighteen century only between 1 and 1.5 million natives remained (Hemming, 1987: 5). In addition to the diseases brought by Europeans, a large number were killed by settlers who waged war on hostile Indian tribes in order to obtain slaves.

The true integration of Amazônia into the capitalist world-economy did not occur until the mid-19th century. In 1839 Charles Goodyear discovered how to vulcanize or harden rubber without destroying its resistance to water. In 1845 R. W. Thompson patented the pneumatic wheel in England and in 1888 John Boyd made the first detachable pneumatic rubber tire (Hemming, 1987: 273). These inventions led to increased demand for rubber, and until 1910 Amazônia was virtually the only supplier. The demand changed the economic, social, and political landscape of the region.

As demand for rubber increased in the international markets, more labor was needed in the region than the local population could supply.

Poor peasants from the Northeast of Brazil, especially from the drought-stricken state of Ceará, were recruited in large numbers to work gathering latex. They joined the bottom of a pyramid dominated by wealthy rubber barons, or *seringalistas*. It has been estimated that by 1850 rubber employed 5,300 men. This figure jumped to 15,400 by 1860 and to 31,000 by 1870. For 1880, 1890, 1900 and 1910 the figures were, 47,300; 82,800; 124,300; and 175,800 respectively. This migration substantially increased the population of Amazônia, from 129,000 in 1840 to 278,250 in 1860, and to 695,112 in 1890. By 1910 there were 1,217,024 people living in the region (Bunker, 1980:18; Hemming, 1977: 273-274). The exact number of northeastern migrants in this population is unknown. Three estimates put them at 117,125 for the period 1872-1900, 160,125 for the period 1877-1900; and 500,000 for the rubber boom as a whole (1879-1920) (Bakx, 1988:144). This rubber cycle began to decline when the British succeeded in adapting rubber trees (*Havea brasiliensis*) to their colonies in Southeast Asia. In less than a decade, Amazônia's share of the market declined from 100% to less than 2%. As a result, per capita income declined sharply, from U.S. $323 in 1910 to U.S. $68 in 1915. As the price of rubber plummeted there was a decline in the population of the region as a whole. In 1910, when the price of rubber in the Belém market was 10,050 reis per kilo, the population was 1,217,024. When the price of rubber declined to 3,570 reis per kilo in 1915 the population also declined to 1,151,548, a 5.4% drop. By 1920 the population had dropped another 5.3% to 1,090,545 (Bunker, 1980:18-19). However, the impact was more dramatic for the more isolated states in the region, such as Acre, where it is thought that at least 25% of the population left the state (Bakx, 1988:149-150).

The decline of the rubber economy meant that people previously involved in the extraction of latex switched to other forms of extraction such as brazilnuts or to subsistence agriculture, often producing for the urban centers of Manaus and Belém (Bakx, 1988:149). But, rubber tapping did not disappear entirely as an economic activity; it survives to the present and rubber tappers have become a strong voice calling for the preservation of the forest on which they depend. During WWII there was renewed demand for Amazonian rubber, and latex extraction flourished as a result of U.S. aid. However, the end of the war again decreased prices and returned the rural economy to its depressed condition (Bunker, 1980: 34; Bakx, 1988:150). Thus, the rubber economy did not provide stable or permanent economic development for most of the region. With the exception of cities such as Belém and Manaus most of Amazônia remained virgin forest, the Brazilian "frontier" waiting to be integrated into the rest of the country.

Manifest Destiny and Nationalism

It has been argued that the image of the future that prevails in a society directs the course of development of that society. This is well-illustrated in the case of Israel: "The resurrection of Israel as preached by the Jewish prophets has been one of the most powerful and persistent images of the future ever evolved, as has the Kingdom of Heaven proclaimed by Jesus" (Polak, 1961: 38). An optimistic image of the future can help mobilize human resources, "for the benefit of the country," and also bring legitimacy to the action of governments.[1]

Amazônia has played a significant role in Brazil's image of the future. Similar to Americans' 19th century perception of the West, Brazilians perceive the Amazon rainforest as an empty wilderness in need of civilization. To put the sparseness of its population into perspective, the area legally defined as the Amazon rain forest comprises 4,900,000 Km^2, 59% of the Brazilian territory. By the 1970's only 3.7% of the Brazilian population lived in this area in a ratio of 1.5 inhabitants per Km^2 (Cardoso and Muller, 1978: 25; de Arruda, 1981: 1). Amazônia has always been perceived as a demographic emptiness, which needed to be conquered and integrated into the rest of Brazil. Unlike its counterpart in the United States, however, Brazilian "manifest destiny" has been devoid of the self-righteous, religious belief in expansion as a God-given right.

Brazilians have viewed expanding "west" as a necessary step for the future development and integration of their country. There has been an awareness that the bulk of the population is located in major coastal cities such as Belo Horizonte, Salvador, Rio de Janeiro, and São Paulo, while most of the country remains "empty," a demographic vacuum as it was referred to by Brazilian scholars and politicians in the 1960's and 1970's. This was seen as dangerous by many of Brazil's leaders, especially in the military, because the emptiness of Amazônia posed a threat to national security. Surrounding foreign powers they feared, and even the United States, would be tempted by the wealth waiting to be discovered in Amazônia and take it over. Colonization by "real Brazilians," would then guarantee de facto ownership of the region. But, colonization was important for another reason as well. It was believed that the region's wealth could fuel the development of the country. It would make Brazil a developed country. Amazonian soils, for example, were believed to be fertile for agriculture because they supported such luxuriant forest vegetation.[2] Thus, Amazônia has been a key element of the "image of the future" Brazilians have had of their country. If it only could be populated, the resources it contained would make Brazil a rich country. This "Manifest Destiny" meant that despite the decline of the rubber industry Amazônia would continue to play an important role in the future development plans of Brazilian leaders.

The dreams of national integration were legally embodied in the Brazilian Constitution of February 24, 1891. The third article of the Constitution mandated the transfer of the national capital from the coast to the Center-West region of Brazil. This Constitution was heavily influenced by the ideals of European positivism, which was accepted in Brazil throughout the course of the nineteenth century, especially in the latter half among military cadets (Skidmore, 1974: 10-11). The writings of Charles Darwin, Herbert Spencer, and especially those of Auguste Comte found in members of the Brazilian elite an interested audience; particularly Comte's views that social progress was attainable through the application of scientific principles in all aspects of society. It played a substantial role in the transition of Brazil from a Monarchy into a Republic in 1889. It provided a new critical mentality, a vision of modern Brazil: a Republic in which science and industry would lead it into a brighter future. For the Brazilian elites, development, e.g., railroads and industry, became a sign of progress. In line with Social Darwinism, they viewed Brazil as a backward country, populated by the assumed inferior African and Indian races. The hope for Brazil was the immigration of the assumed racially superior Europeans and the adoption of their culture (Burns, 1980: 207-208). This view of progress became associated with the positivistic notion of order; that is, social order or social stability as necessary for development. Thus "order and progress" became the national motto, the symbol of the Brazil to be built. These words, *ordem e progresso*, are still written on the Brazilian flag.

Amazônia was part of this positivistic vision of progress. It was hoped that the transfer of the capital from Rio de Janeiro to the Center-West would act as a magnet for Brazil's surplus population then concentrated in the cities of the coast, thus developing the empty interior of the country (Kubitschek, 1975: 8, 20). However, the mandate of the (Republican) Constitution of 1891 was burdensome and difficult to carry out, due to the costs involved in building an entire city and relocating the federal government. Consecutive Brazilian presidents simply avoided it, though on September 7th, 1922 a marker was symbolically placed where Brasília was to be built.

Brazil entered the 20th century as a backward peripheral country, producing mostly coffee and sugar for export to European and North American markets. Right after independence from Portugal in 1821, Brazil became essentially a periphery of Great Britain, thanks to trade agreements that Britain demanded if it was to recognize Brazil as an independent country. British commodities flooded the Brazilian market throughout the nineteenth century, thus preventing the emergence of a Brazilian industrial apparatus (Barbosa and Hall 1986; Barbosa, 1989; Manchester, 1933).[3]

In the first half of the 20th century, however, economic nationalism became a driving force for industrialization of the country. This nationalism maintained the goals of modernization inherited from positivism while addressing the problem of economic dependence neglected by positivists. It had the intellectual backing of scholars such as Alberto Torres who wrote *O Problema Nacional Brasileiro* (1914), The Brazilian National Problem. In that book he argued that foreign companies were taking control of the Brazilian economy and urged the government to do something to stop it (see pp. 95-96). Beginning with President Getúlio Vargas in 1930, industrialization was defined as the key to economic independence. Industrialization combined with national control of natural resources and the building of an infrastructure of roads and railroads would determine the future of Brazil as a world power. Considering its geographical size and its natural resources, Brazil was bound to become an industrialized nation. Government projects such as the nationalization of oil, the building of smelters, roads, etc., stirred a great deal of nationalism, and often anti-imperialist sentiments.

It was not until the presidency of Juscelino Kubitschek (1956-1961) that the transfer of the capital, as mandated by the Constitution of 1891, became reality.[4] President Kubitschek had an urgent vision of development for Brazil, one filled with nationalism and pride. For him Brazil would have to advance fifty years in five in order to catch up with the developed countries (Viola, 1988: 212). And, the development of the interior would play a key role in this process. There the new capital, Brasília, would be built. He provides the following account in his book *Por que Construí Brasília?* (1975) (Why I Built Brasília):

I had seen Brazil from above--from inside an airplane--and I could see the problem in all its complexities. Two thirds of the national territory lacked human presence. It was the "demographic vacuums" of which sociologists spoke. The greatest challenge of our history was there: to force a shift in the axis of national development. Instead of the littoral--which had already achieved a certain level of progress-- the Central Plateau would be populated (p. 8).[5]

The new capital would act as a magnate, a pole of development that would expand Brazilian civilization to the interior:

> The population nucleus created in that faraway region would spread like an oil spill, making all the interior open its eyes to a glorious future for the country. Thus, the Brazilian would be able to take possession of his vast territory. And the change of the capital would be the vehicle, the instrument, the factor, that would make this new cycle of conquest [*ciclo da bandeira*] possible (pp. 8-9).[6]

The economic integration of the country and the development of infrastructure were part of a bloodless revolution to change the course of Brazilian history:

> For this objective to be achieved a revolution would have to take place, not a bloody revolution but one of administrative methods. First, Brazil would have to eliminate its empty spaces. For this to happen several things would have to take place: setting in motion the exploration of [Brazil's] immense natural resources; ending its enormous social inequalities; bringing closer the population centers by opening roads in all directions; providing abundant and cheap electricity to the states; building hydroelectric plants where they are needed; attracting foreign capital to make possible the construction of smelters aiming at national industrialization; irrigating the Northeast in order to stimulate its agriculture; laying open the Amazon forest in order to incorporate it to the rest of the national territory; and finally, moving government headquarters by building the new capital in the geographic center of the country (p. 13).[7]

Kubitschek wanted to go beyond the building of a new capital per se. He envisioned Brasília as a true center of integration for the country as a whole. It was supposed to be the center of a network of roads which would join "the gauchos of the pampas to the rubber tappers of the Amazon. . ." (Kubitschek, 1975: 8-9). These roads would allow the population center created by Brasília to spread "like an oil spill, awakening the interior to the great future of the country" (Kubitschek, 1975: 8-9). Even before the inauguration of Brasília on April 21, 1960, the workers building from both ends of the Belém-Brasília highway, connecting Brasília to the city of Belém do Pará in the northern State of Pará, met in 1959. On July 4, 1960 the workers from both ends of the Brasília-Acre highway also met. President Kubitschek recalled the importance of this road by making parallels between Brazil and the United States:

> In 1850 the United States had laid open immense western pastures, aiming to link the Atlantic to the Pacific. When I initiated the Brasília-Acre [highway], I was realizing an identical adventure. And, I was doing it with only a one-century delay. The objective of the road was the immediate integration of the southwest region of Amazônia (Kubitschek, 1975: 314).[8]

Brasília was also connected to the major coastal cities of Fortaleza in the Northeast, Belo Horizonte in the state of Minas Gerais, Rio de Janeiro, São Paulo, and other cities in the South. However, it was the construction of the Belém-Brasília and the Brasília-Acre highways that would pave the way for the later invasion of the Amazon rain forest since these roads cut directly through it (Burns, 1980: 460; Kubitschek, 1974: 8-9).

Brasília became the symbol of the Kubitschek administration's developmentalist nationalism, a symbol of Brazil's supposed "destiny" to undertake as fast as possible a policy of development. It represented the opening of land and resources available for the building of the new Brazil, the 20th century Brazil (Skidmore, 1967: 167-168). It would allow for the filling of the "empty space" of the forest. It would also open the doors for solutions to major social problems pressing Brazilian society by providing the landless masses, especially the poor of Northeast Brazil, with land (Cardoso and Muller, 1978: 141). And it did work. It triggered the expected population growth. The city was originally planned to accommodate 500,000 people. By 1996, however, its population plus the surrounding areas was over 1.8 million people (Civila, n.d.; *O Estado de São Paulo, NetEstado*, 1997). The Center-West region, where it is located, experienced a positive net migration of 23.22% between 1960 and 1970. However, Kubitschek's "oil spill" on the "empty" forest of the North was much more modest, even though the Center-West state of Mato Grosso, part of the Amazon region, had a net migration of 27.38%. The North region, where the bulk of Amazon rainforest is located, experienced a net migration of only 2.68% between 1960 and 1970, and this positive migration was uneven. While the state of Pará, for example, had a positive net migration of 5.52%, the state of Amazonas, the largest rain forest state, experienced a net migration of -2.40%, the state of Maranhão -8.85%, and the territory of Acre -2.30% (Graham and de Hollanda Filho, 1971: 22-23;

see also Mougeot, 1985: 61). It was not until the 1980's that migrations to certain areas of Amazônia, such as Acre and Amapá, achieved high levels. These migrations followed the building of roads to these areas. It is no coincidence that the area most affected by deforestation in Amazônia in the 1970's and 1980's was along the Belém-Brasília highway. As the military improved the conditions of these roads and new ones were opened, population followed. In 1968 the 1,500 km Cuibá-Pôrto Velho highway (BR-364) opened up Rondônia for settlement. As a result, the average migratory flow to that state (a federal territory when the road was opened) increased tenfold in the 1970's. Whereas it was approximately 3,000 per year in the 1960's, when the road was paved in 1984, about 160,000 migrants a year entered Rondônia (Mahar, 1990: 63-64).

The building of Brasília marks the beginning of the "planned" preda-tory exploitation of the Amazon rain forest. The colonization and utilization of the resources of Brazil's "empty space" would thereafter involve the full participation of the state. When the military took over the government in 1964 this process of planned invasion intensified. Kubitschek's ideology of rapid economic growth and development as a sign of progress seemed actualized (Viola, 1988: 213). This time, however, the ideology became associated with the recognition of Brazil as a "future superpower" and the need to protect Brazil against possible foreign invasions.

Brazilian Insecurities over Amazônia

Brazilian nationalism about Amazônia needs to be understood from the perspective of Brazil's expansionist history in South America, and its own imperialistic tendencies in the region. A quick examination of any historical atlas reveals that from its discovery in 1500, Brazil has increasingly expanded westward. When the Pope divided the New World between Spain and Portugal with the Treaty of Tordesillas in 1493, the area accorded to Portugal in South America was much smaller than the area occupied by Brazil today (see Buescu, 1970:51). The current size of the country is the result of claims made by *bandeirante*[9] expeditions and de facto occupation. However, these claims always left Brazil insecure due to the small population of the interior. A foreign takeover was perceived as a possibility.

Brazilian imperialism in the Amazon region is well illustrated by the acquisition of the territory of Acre from Bolivia. During the rubber boom of the late nineteenth century, Brazilian rubber barons and tappers moved into territory which was then part of Bolivia. The wealth generated by rubber induced a border dispute. The conflict involved secret deals between the Bolivian government and the United States, which had a vested interest in the rubber being produced in the region.[10]

Frustrated by the inability of Brazil and Bolivia to reach an agreement on their future, people in the region declared their independence from both countries, creating the independent State of Acre, also known as the Republic of Poets on July 14, 1899. The newly independent state, composed primarily of Brazilians, managed to expel the Bolivians from the area, thus opening the door for Brazil's de facto control. This goal was achieved diplomatically in the Treaty of Petrópolis of November 17, 1903, when Brazil purchased Acre from Bolivia for a sum of two million U.S. dollars. In signing this treaty, "The Bolivians bowed to the inevitable" (Hecht and Cockburn, 1989:72). Despite the fact that Brazil was the imperialist power in this case, the involvement of powers such as the United States and the possibility of Bolivia reclaiming the region for itself, have played on Brazilian insecurities. Other incidents played on these insecurities. For example, a casual suggestion made in a speech by Harrison Brown, author of *The Challenge of Man's Future*, that the surplus population of India could be moved to the Amazon region made headlines in Brazil (Fearnside, 1986:17). Other similar statements concerning the region aroused the possibility of a takeover for Brazilians.

These fears of losing Amazônia to foreign powers are summed up in a xenophobic essay by Arthur Cézar Ferreira Reis entitled *A Amazônia e a Cobiça Internacional (Amazônia and International Covetousness)*. The 1982 edition was the fifth publication of this book since it was originally published in 1957 (see Fearnside, 1986:17 also). The xenophobia of this book borders on paranoia. In the preface to the second edition the author criticizes the participation of scientific ventures in Amazônia involving American institutions such as the Academy of Science. He sees the involvement of these organizations as part of a hidden plot to take over the region (see Ferreira Reis, 1982:4-6).

The fear of foreign takeover expressed by Ferreira Reis was also prevalent at an important Brazilian military institution, the Escola Superior de Guerra, or the Superior War College. One of its most famous graduates, General Golbery do Couto e Silva became one of the leading proponents of populating and integrating Amazônia with the rest of Brazil (Zirker and Henberg, 1994:270). His view was that Brazilian development policy should follow the geopolical needs of the country; that is, of a country whose population was concentrated primarily in the littoral and with a population vacuum in the interior. For him, the government had to formulate a "grand strategy" of development taking into account the long-term needs of the country. The development of Amazônia would create a protection against the emergence of communist guerrillas and foreign takeover. *Operation Amazônia*, the military's scheme to develop the region, became the practical expression of Golbery's "grand scheme" for Amazônia (Hecht and Cockburn, 1989:103-104).

National Security and "Brazil: a Superpower in the Year 2000"

From the military's perspective the purpose of the coup of March 31, 1964 was to "restore" order by taking the government back from (those perceived as) "communists;" that is, the political left headed by President João Goulart (1961-1964) (see Burns, 1980: 509). Goulart was a populist president with strong ties to labor and peasants. He was also attempting to establish closer connections between Brazil and the communist bloc, in Cuba and China. This was too much for the military who ousted him.

The military takeover brought to the country an ideology of great nationalism. Patriotism, as defined by the military, was in and any dissension was out. For Amazônia this nationalism would be translated into "National Security." Since Brazil shares borders with all countries in South America except Chile and Ecuador, the Amazonian "demographic vacuum" (*vazio demográfico*) could be invaded at any time. By populating the frontier, the country would be secured against a takeover (see de Arruda, 1981 for an argument along these lines).[11] The fear of a takeover was exacerbated by the Indian problem. The territory of the Yanomami Indians, for example, lies both in Brazil and Venezuela; it was believed that in case of an invasion the Yanomamis could side with the enemy. Also, in the eyes of the military there was always the possibility that Indians would declare independence and form small states. Another perceived threat were the guerrillas who, opposing the military, escaped into the jungle (see *Veja*, October 13, 1993, pp.23-28). It was thought that Amazônia would become a harbor for guerilla activity. There was also the possibility of infiltration by the socialist government of Surinam in the northeastern Amazônia (Goldemberg and Durham, 1990, 30). The fear and the rationale for developing Amazônia is illustrated in the following quote:

Amazonia must be traversed by roads to provide infrastructure and facilitate settlement but also "for easier defense against invasion," a general told me in 1972 (as if unbroken forest is not a better defense?), "and against insurgents" (Guppy, 1984: 940).

National Security has been the focus of military policy concerning Amazônia since the 1960's.   Even with environmentalist pressure mounting against Brazil in the 1980's they held to that policy.   José Goldemberg and Eunice Ribeiro Durham (1990:30) argue that "This doctrine [National Security] was created by the United States to insure its hegemony in Latin America and was adopted by large sectors of Brazil's armed forces."   According to them, it led to the "militarization" of the region.

The quickest way to populate Amazônia was to make use of the surplus, landless population of other regions of the country, especially the Northeast.   By populating Amazônia with landless peasants the government could, of course, also use the forest as a safety valve against demographic unrest.   It would appease the peasants by providing them with land, and appease the landowners by removing the threat of land invasion. This was certainly a more politically prudent enterprise than land reform, in a country where the rural areas have been dominated by a few powerful landowners who also provided the military its base of support.[12]

Populating the Amazon rain forest would also allow for the exploitation of needed resources to fuel the Brazilian economy, and thus revitalize Brazil's participation in the world economy.   Amazônia would provide many of the raw materials necessary to make "Brazil a Superpower in the year 2000," a slogan promoted during the military years.   This slogan symbolizes an optimism resulting from the high economic indicators of the early 1970's, when Brazil achieved a growth rate of 11% á year, along with the belief that Brazil had enough natural resources and empty spaces to sustain this development.   It was the Brazilian "miracle" of quick development.   The social inequality and environmental problems created by such development were issues of secondary importance (see Burns, 1980; and Roett, 1984).

In order to coordinate development the military created a series of agencies responsible for different regions of the country.   For Amazônia, the main agency was the Superintendence for the Development of Amazônia (SUDAM).   But other key agencies were created to handle different aspects of development.   The Institute for Colonization and Agrarian Reform (INCRA) was responsible for the relocation of peasants to the region (Bakx, 1988: 152-154).   The scientific exploration for mineral deposits was dispersed among several agencies, including, the National Institute of Amazonian Research (INPA).   Perhaps the most significant effort was Project RADAM (Radar in Amazônia). In addition to mapping the area through aerial photographs, it also catalogued resources, e.g., mineral deposits, and soil (de Arruda, 1981: 1-2).

The responsibility for forest protection was under the Brazilian Institute of Forest Development (IBDF),[13] and Indian affairs under the National Indian Foundation (FUNAI). These two agencies differed from the others in their emphasis on protection rather then exploitation. However, this difference soon became de jure. Both the IBDF and FUNAI often complied with the interests of government and private capital (Bunker, 1985: 116-119). When powerful interests demanded the clearing of forests or the removal of Indians these organizations approved or looked the other way. And since all contact with the Indians had to be made through FUNAI, critics feared that this policy was an attempt to prevent outsiders from knowing what was really happening to the Indians (Comitê Interdisciplinar de Estudos sobre Calha Norte, 1988:74).

In line with the military mentality of the period, especially as it concerns the concept of national security, the "development" of Amazônia was conducted in a semi-military fashion. The effort was called "Operation Amazônia" (Hecht, 1985: 670). The development was to occur under the administration of a group of civilian institutions operating under the guidance of the military government, which was anxious to develop a region it perceived as vital for national security and the future of the country.

It was difficult to mount opposition to this military vision of development for Amazônia (Viola, 1988: 213). Through a series of dictatorial "institutional acts," successive military governments deprived "undesirable" politicians of their political rights. Article X of Institutional Act No. 1, of April 9, 1964, for example, gave the president the right to revoke legislative mandates and to suspend political rights (Roett, 1984:128). Individuals who opposed the government were often accused of having *idéias alienígenas* (alien ideas), and labeled as "communists" (and were thus subject to incarceration) regardless of their true political position. These policies halted civil disobedience, not only by environmentalists but by all groups (see Fontaine, 1985 for the Antiracism Movement). During the 1960's and 1970's the governments's slogan was "Brazil: Love It or Leave It." Those who protested should either leave or be put in jail. Open opposition to the system from small urban guerrilla groups that kidnapped foreign envoys such as the American, West Geman, and Swiss ambassadors, and a Japanese consul, in the late 1960's and early 1970's was crushed (see *The New York Times*, September 5, 1969 for the American ambassador; June 12 1970, for the West German ambassador, December 8, 1970, for the Swiss ambassador, March 11, 1971 for the Japanese Consul).

While the overall goal of populating Amazônia was unchanged and unchallenged during military rule, plans for doing so changed substantially. The government began by providing fiscal incentives for investment in Amazônia, then created a plan based on colonization by small producers, and finally a plan of development based on large projects. Each of these strategies was a change of emphasis in response to specific obstacles in the development process itself, or elsewhere in Brazil. The program of allocating small properties to landless peasants, for example, did not cease when large landowners replaced small ones in government policies, nor did economic incentives for investments.

The economic incentives for investments in the Amazon region began with Law 5.1744 of October, 1966. This law allowed a percentage (initially 50%) of a corporation's tax liability to be invested in projects in Amazônia. This essentially allowed taxes owed to become venture capital. In addition, the government established easy credit which greatly facilitated the development of new projects. Thanks to these incentives, projects approved by the government increased substantially, especially those for cattle ranching. The period of peak investments was from 1967 to 1972 (Bunker, 1985: 85; Hecht, 1985: 670-671). Investors obtained property rights for land that they cleared simply to get tax benefits. By 1979 the government, realizing that many of the projects were only schemes to avoid taxes, reduced the incentives (Hall, 1987:531). The incentives, coupled with the requirement that the owner "improve the land," that is clear it, led to the destruction of vast tracks of virgin forest. Much of the forest cleared for cattle ranching eventually became unproductive. After a period of five years the soil became too impoverished to sustain the required pasture. So cattle ranchers simply moved further on by opening more virgin forest (Fearnside, 1988a:284-285).

The period from 1970 to 1974 marked an interlude in a policy favoring big business to one that centered on the idea of using the Amazon as a settlement frontier for the landless poor of Northeast Brazil. The project came under the ideological banner of "*O Homem é a Meta*" (Man is the Goal), spelled out in the National Development Plan I (PND I). In this period the building of the Transamazon Highway began, under the slogan that it would connect "land without men with men without land" (Bunker, 1985: 111; Hecht, 1985: 672; Velho, 1984: 35). In order to generate a migration of landless people the government developed towns, e.g., Marabá, Altamira, Humaitá, and Itaituba, along the highway to attract them.[14] These development towns failed to attract the number of expected colonists. By 1974, the towns of Marabá, Altamira, and Itaituba had taken only between six and seven thousand families (Mougeot, 1985: 67). Even so, this should not be seen as a failure of the Amazon region as a whole to attract migrants. From 1960 to 1970 the "new frontier" (Rondônia, Acre, Amazonas, Roraima, Pará, Amapá, and Mato Grosso) had a total rural population of 164,669 people. From 1970 to 1980 this figure increased to 1,047,912 (Velho, 1984: 39). The problem, however, was that many of these individuals continued to be landless, working as migrant workers in the region. Many gave up trying to earn a living in the countryside and migrated to the larger cities of the region, i.e., Manaus and Belém. From 1970 to 1980 the urban population of Northern Brazil increased from 1,626,600 in 1970 to 3,046,129 (Benchimol, 1985: 49). This process of migration continues at a fast rate today. The development town of Marabá itself is one of the fastest growing in Brazil with a 17.5% population growth in the 1980's along with severe problems of poverty and disease (e.g., 3,200 cases of malaria, 150 cases of tuberculosis). It grew from 49,000 inhabitants in the early 1980's to 270,000 by the end of the decade. Two other Amazonian towns are also among the fastest growing: Ariquemes (in Roraima), 13.5%, and Pôrto Velho (in Rondônia), 8.5% (*Jornal do Brasil*, July 2, 1990).[15]

From 1975 to 1979 the government shifted policies to large land grants under the second National Development Plan (PND II). The agency responsible for the general development of the Amazon, SUDAM, argued that the development of Amazônia based on small properties was handing the land, "to ignorant people," and causing deforestation and the exhaustion of soil for subsistence agriculture. Large cattle enterprises for export were promoted as the prototype for development. These ranches were described as environmentally rational. There was a perception, based on no scientific evidence, that cattle ranching actually improved the quality of the soil by increasing soil nutrients, and therefore represented an "environmentally sound enterprise." Roughly 10 million hectares of land were thus converted from forest to pasture in Amazônia (Hecht, 1985: 673). As already pointed out, the soils of Amazônia cannot sustain cattle ranching for long due to declining nutrients. The emphasis on large cattle projects meant a shift from one unsustainable form of colonization to another. In addition to cattle ranching, however, whole areas of Amazônia were also set aside to become poles of development, experiments in capitalist development in a fragile tropical ecosystem. These mega projects were to be implemented in partnership with international capital.

One of the ironic features of the Brazilian military dictatorship was that while the military's rhetoric was to a large extent xenophobic, especially as it concerns the Amazon rainforest and the fear that other countries would eventually invade it, foreign capital actually sustained the government. The military coup had the blessings of the United States which feared that Brazil was turning into another Cuba. The reward for the United States was that

> The coup d'etat of 1964 resulted in the reversal of the trend toward economic nationalism and in a particularly strong repudiation of the "independent" foreign policy. One of the first acts of the new government was the revocation of the law limiting the remittance of profits by foreign firms. That act, with the opening of various areas of the national economy to foreign investments and the signing of a bilateral investment guarantee treaty, resulted in a sharply increased flow of private United States capital to Brazil (Black, 1977:49).

U.S. investment in Brazil increased from $846 million in 1966 to $2.033 billion in 1971 (Evans, 1979:168; see also Roett, 1984: 169). Also, U.S. aid to Brazil constituted "the third [largest] program of U.S. 'assistance' in the world (following U.S. aid to Vietnam and India)" (Black, 1977: 61). Despite this "partnership" with foreign capital, protectionism fueled by nationalism survived in certain sectors of the economy. Petroleum exploration and ownership and telecommunications, for example, remained monopolies of the government. These were viewed as of "national interest," and thus remained in the public sector in an attempt to prevent the total control of the Brazilian economy by foreign interests. But foreign capital dominated major economic sectors, e.g., by 1971, 62.7% of Brazil's metallurgy industry was controlled by foreign firms, 68.4% of the machinery and 64.9% of the electric and communication equipment industries as well (Roett, 1984:168).

The Miracle

The combination of public and private capital, cheap oil prices, and loans from international organizations led to high rates of economic expansion. In the late 1960's and early 1970's the rate of economic growth reached its apex. The industrial sector and the manufacturing industries expanded at rates of 13.2% and 13.9% respectively while GDP grew at average annual rates of 11.5%. In this period Brazil became the darling of international development organizations. They pointed to the Brazilian experience of development as the model for other developing countries, much as the Asian model of development was praised in the 1980's and 1990's. A major problem with the Brazilian miracle, however, is that it was so dependent not only on foreign capital but on the availability of cheap oil in the world economy. When OPEC increased oil prices in the early 1970's the miracle was sustained only by the availability of petrodollars. By the end of 1977 Brazil's net debt had risen to $32 billion and debt servicing required 51.2% of that year's exports (Roett, 1984: 168-170). High economic performance imposed high future costs. The country continued to borrow heavily, averaging U.S.$17 billion each year. By the early 1980's Brazil was in the middle of a debt crisis, threatening its creditors with default.

What is tragic about the Brazilian "miracle" is that while the economy achieved high indicators the bulk of the population remained poor or got poorer. The miracle was for the elites, who lived a lavish lifestyle, "eating cake," while the poor gained very little or went hungry. In the major cities the poor lived in squalid slums surrounding the neighborhoods of the rich, often providing them with the cheap labor in factories and in service. The miracle was especially bad for people in the countryside who lost their livelihood due to the mechanization of agriculture, and the consolidation of land for production for the international markets, etc. These are the landless peasants for whom the Amazon rainforest became a destination.

With the rise of the military, the participation of foreign capital grew in importance (see Campuzano, 1979; and Irwin,1977). The first projects in Amazônia were a gamble because they involved economic activities in agriculture and cattle ranching in a region where the success of these activities was uncertain. Thus, they were largely left to the private sector which took advantage of tax incentives. As huge mineral deposits were found, e.g., iron in the Carajás region, it was easier for the Brazilian government to forge economic alliances with multinational capital and obtain loans from international institutions such as the World Bank.

Notes

[1.] A pessimistic view of the future is, of course, also possible.

[2.] This has proved to be an incorrect assumption. The soils of Amazônia are primarily Oxisols (45%) and Ultisols (29.4%) which are not very conducive to agriculture (Furley, 1990:312-314), the forest supporting itself through an extremely efficient recycling system where dead matter is quickly transformed by living species.

[3.] Britain obtained the Brazilian market as a result of the help it provided the Portuguese Royal Family in escaping Napoleonic forces invading Portugal in 1807. As a result of this help the British commodities received a preferential treatment in Brazilian ports. While Portuguese commodities paid 16% ad valorem, British commodities paid only 15% (Barbosa, 1989: Ch. IV).

[4.] In his semi-autobiographical book *Por que Construí Brasília?* (Why I built Brasília) (1975) President Kubitschek wrote that his commitment to the cause of building Brasília happened almost by accident. At a political rally he was asked by a journalist if he was going to implement that constitutional requirement. Caught by surprise, he replied, "yes."

[5.] Author's translation.

[6.] Author's translation.

[7.] Author's translation .

[8.] Author's translation.

[9.] *Bandeirantes* were Brazilian explorers who claimed the areas they found for Brazil.

[10.] For example, 'In March of 1899, the *Wilmington* of the US Navy tied up at Pará [in Brazil], and [Lt.] Chapman Todd stepped ashore to render greetings to the governor there. He spoke eloquently of the "love which dignifies the human species" but his real mission was to carry a secret agreement from Paravacini, the Bolivian diplomat eagerly awaiting him, to the US government' (Hecht and Cockburn, 1989:70).

[11.] This view of national security is maintained even today by the military. The ecological movement in Brazil is often accused by them of disregarding the security of the nation. Preservation in this view, then, means that the Brazilian frontier will remain open to foreign invasion (*Jornal do Brasil*, July 12, 1990)

[12.] The success of colonization of the Amazon rainforest as a solution to land disputes in the rest of Brazil is yet to be seen, however.

[13.] IBDF no longer exists. It was absorbed by the Brazilian Environmental Institute (IBAMA) in 1988.

[14.] These projects were known as Integrated Colonization Projects (Projetos Integrados de Colonizacão, PIC) (Bunker, 1985: 11).

[15.] These figures were reported by the Brazilian Institute of Geography and Statistics (IBGE).

Chapter 3

From Ranches to Pyramids: The Changing Nature of Projects in Amazônia

One important characteristic of development projects in Amazônia is that as time progressed their size increased. Initially, privately-owned ranches were large, especially when compared to the plots of land owned by poor peasants. They also led to much deforestation. However, in comparison to later projects, their size and impact were lilliputian. Projects such as Calha Norte and Carajás cover 14% and 10.6% of the Brazilian national territory, respectively. These projects, were devised without public debate. The few voices of dissent suffered lack of coverage by a suppressed media quickly silenced by the dictatorship. Dissent increased somewhat after the end of the military regimes, but by then implementation was already in full swing.

Brazilians call the huge projects that have been devised by their politicians *obras faraônicas* (pharaonic undertakings) and *projetos faraônicos* (pharaonic projects). Like Egyptian pharaohs, politicians have spent millions of dollars in huge, costly public works with little benefit for the people. Perhaps, like Egyptian pharaohs, the intention was to leave their names for posterity. The common expression *faraônico(ca)* reveals the popular attitude towards Brazilian modernization plans from the time of President Juscelino Kubitschek, with the construction of Brasília, on. *Projetos faraônicos* were a major feature of the military period; the rulers claimed that a big country required big projects. The Itaipu Dam in southern Brazil, the Rio-Niterói Bridge, connecting both sides of Guanabara Bay, and the Transamazon highway, and the Jari, Carajás, and Calha Norte Projects in Amazônia were massive projects by world standard. Undertakings of this kind in Amazônia reflect a time of modernization with little concern for the environment and the fate of the people of the forest. They were an orchestrated invasion of Amazônia by capitalist forces. We begin with the smaller, privately-owned cattle projects and then address the major projects.

Arm-Chair Cowboys

Road building in Amazônia allowed cattle ranching to spread very quickly. This was especially true in the states of Mato Grosso and Pará where the building of roads such as that between Belém and Brasília brought a flux of migrants and the development of major cattle projects. An estimated 90,000 km^2 were cleared in Amazônia in the 1970's for cattle ranching alone (Kohhepp, 1981:77; and Branford and Glock, 1985:20). These first cattle projects were placed in an environment that people understood very little.

Who owned these ranches? Who would be interested in such ventures? The owners were either wealthy Brazilian private and corporate owners from the developed southeast region of the country or foreign enterprises, often a combination of both types (see Campuzano, 1979; and Irwin, 1977). They were attracted by land at low prices and the generous tax rebates given by the Brazilian government. Between 1966 and 1976 Suiá-Missu received tax rebates worth 21.4 million British pounds. It was owned by a wealthy Brazilian family, the Omettos, who eventually sold it to an Italian corporation, Liguigás. Two other ranches, the 180,000 hectare Codeara and BCN Agropastoril, were owned by the São Paulo based Banco de Crédito Nacional. These ranches failed to generate a profit due to the usual reason: the inability of the soil to sustain cattle. Attempts to diversify into other economic activities, such as sugar cane production, achieved mixed results. In the 1980's foreign capital came to the rescue. In April 1983, Codeara and the Brazilian Goodyear subsidiary formed a joint venture to create a rubber plantation on Codeara's land. The investment of 3.7 million British pounds was at the time the biggest attempt to diversify away from cattle-rearing (Branford and Glock, 1985:109-115).

One would think that the long-term unsustainability of cattle ranching in the region would become a deterrence to the onset of new cattle enterprises. However, profits were made from subsidies and from initially fertile soil. As the soil fertility declined new forest were opened for new pastures. Also the land increased in value due to the development of infrastructure, e.g., towns, nearby and the development of a land market in Amazônia characterized by much speculation (Kohlhepp, 1981:77; Hecht, 1985: 678). This led to a situation in which "...the exchange value of the land itself was far higher in this speculative context. Entrepreneurs depended for profits, *not on the annual productivity of the land, but on the rate of return of investment"* (Hecht, 1985: 678).

In addition to deforestation, the development of cattle ranching led to major land conflicts in the region. Cattle ranches encroached on lands owned by Indians as well as those owned by peasants. The Indians had only ancestral claims to the land and the peasants often had no title. This lack of documentation coupled with the rancher's financial means to hire gunmen gave cattle ranchers the upper hand. In areas like the south of Pará and the North of Mato Grosso, where cattle ranching became the dominant economic activity, 6.7% of the landholders controlled fully 85% of the private sector lands while 70% of the farmers controlled a meager 6% (Hecht, 1985: 679).

According to government figures, 38% of all deforestation that took place in Amazônia between 1966 and 1975 was attributed to large-scale cattle ranching (agriculture contributed 31% and highway construction 27%). It is important to keep in mind that "The government gave fiscal incentives to 90 percent of the ranches, and more than half the agricultural clearing was under a government-sponsored peasant colonization program" (Caufield, 1991:40).

Project Jari

By any standards Jari was a huge project, a *projeto faraônico*, a pyramid, and more than any of the cattle and farming projects before it set the size of the projects that followed. Jari became a 1.6 million hectare tree and rice farm at the confluence of the Amazon and Jari rivers in the state of Pará (Gaspari, 1998). The area was bought at the low price of 75 cents an acre by American tycoon Daniel Ludwig. Foreseeing a shortage of paper created by increased demand brought about by literacy efforts in the Third World, Ludwig's plan was to plant trees and process them into paper pulp. Wanting full control of Jari in the beginning, he refused economic help from the Brazilian government. The deal led President Castello Branco to express the hope that "Brazilians would follow Ludwig's lead in opening the Amazon" (Kinkaid, 1981:105). Later, the military viewed Ludwig's Jari with suspicion, as a foreign threat to Brazil's national sovereignty over Amazônia.

The clearing of virgin forest for Jari was done by burning the jungle. There were so many fires that thunderstorms were created miles away. The fragile soil of the area was so compacted by the bulldozers used to clear the forest that when *Gmelina arborea*, a fast-growing tree imported from Asia, was planted, many of the seedlings quickly died. The operation then switched to the use of manual labor by seasonal laborers, but this drove costs up (Kinkaid, 1981:105-106; Branford and Glock, 1985: 79).

Ludwig originally planned to invest between U.S. $300 to $500 million dollars in Jari; when he finally unloaded the project to a Brazilian private consortium in 1982 he had invested U.S.$1.5 billion (Gaspari, 1998). A major reason for his failure was that the soils of Amazônia did not respond as expected. By the early 1980's average yields for *Gmelina arborea* in clay soils were 40% off target, and on the poorer sandy soils 75%. Rice yields were also 30% below target, losing U.S.$8 million to U.S.$10 million a year (Gaspari, 1998; Kinkaid, 1981:110; Fearnside, 1982:326; Arruda, 1979). Despite *Gmelina's* lower than expected yields on February 18, 1976 Ludwig ordered a pulp mill. This mill became the epithet of an *obra faraônica*. It was built in Japan by Ishikawajima-Harima Heavy Industries and floated across three oceans to Jari. It required a U.S.$240-million loan guaranteed by the Brazilian government from the Japanese Export Import Bank. Another U.S.$269-million loan covered the rest of the required financing (Kinkaid, 1981:109).

Ludwig's arrogance about Jari sparked Brazilian nationalism. Until 1979 the press was not allowed in and there were rumors in Brazil that even President João Batista Figueiredo was not allowed to land there for a visit. In a period of transition to democracy when the press was un-gagged, Jari became "a focus for left-and-right-wing fears that imperialist foreigners would expropriate the Amazon riches" (Kinkaid, 1981:109). When Ludwig sought government approval he expected the Brazilian government to guarantee a second Japanese loan for the purchase. He also expected the World Bank to approve loans for a power plant to be constructed for a bauxite smelter. These projects would cost U.S. $700 million dollars. The negotiations dragged on as opposition grew. The government insisted that at least 70% of the mill be made in Brazil, and the project proved too controversial for the World Bank when Ludwig disclosed that he had given Jari to his Zurich Cancer Society. Without World Bank support Ludwig could not borrow from private banks (Gaspari, 1998; Kinkaid, 1981:109-110).

In 1981 with Jari's debt at U.S.$370 million Ludwig decided that he had enough. The project's economic failures forced his Universe Tankships Corporations to unload it to a consortium of 22 Brazilian companies in 1982. It was renamed "New Jari" (Gaspari, 1998; Fearnside, 1988b:83; Kinkaid, 1981:114). Even though it was purchased by a private consortium, in reality the transaction was a facade constructed to maintain Brazil's "good name" among foreign investors at a time when a dwindling economy made such investments necessary. In the end the Brazilian people paid the bill. New Jari made a profit only during one year, 1994. It has been subsidized by the Brazilian government since it was bought. It has been suggested that New Jari may end up costing the Brazilian taxpayer as much as U.S.$205 million (Gaspari, 1998).

TransAmazônia: A "Pyramid" With a Populist Twist

The Northeast of Brazil has been a geographical area plagued with droughts throughout its history. These severe droughts brought with them death and misery to millions of peasants already in fragile living situations. The region is characterized by high levels of inequality in which the majority of the people go hungry or malnourished while a few wealthy landowners control the majority of the land, the "rivers of sugar" (because of the predominance of sugar cane plantations). These landowners are known as *coronéis* or colonels, a fictitious military title fit for authoritarian feudal lords. Politically these individuals have been very powerful in Brazilian history, controlling the vote of the peasants under their control. While the *coronéis* have the standard of living of first-world countries, the poor peasants have the standard of living of slaves.

In the early 1970's the Northeast was experiencing a severe drought. The situation was so severe that it even moved the hearts of dictators. President (General) Emílio Garrastazu Médici visited the region and commented: "I saw with my own eyes . . . . Nothing in my whole life has shocked and upset me so deeply. Never have I faced such a challenge" (in Branford and Glock, 1985:60). Shortly after his visit the National Integration Plan (PIN) was announced. It called for investments in irrigation projects in the Northeast and the construction of the Transamazon highway. The initial capital for the construction was both Brazilian and foreign. PIN had a budget of U.S.$450 million dollars, obtained by tapping 30% of the budget of the agencies responsible for the development of Amazônia. A total of U.S.$100 million dollars were drawn from this fund for the construction of the road. Another U.S.$10 million dollars came from USAID. The World Bank did not contribute directly for the construction of the road; it did, however, provide U.S.$629 million dollars to the Brazilian government between 1970 and 1979 for the construction or improvement of roads. The Bank also made available an additional U.S.$132 million dollars for integrated development projects which included the building and repair of roads. The result was that "These generous allocations of money . . . allowed the Brazilian government to shift road construction funds to Amazônia that would have been spent in other regions" (Smith, 1982:14-15).

Without congressional or public debate, the plan was approved. Thus, instead of dealing with the politically explosive problem of land reform in the Northeast, the government opted to create an "escape valve" for the problem of poverty in the region. Peasants were encouraged to migrate to the Amazon region in search of a better life where they would populate the demographic vacuums of the region, and take de facto control of it for Brazil. The specific plan for Amazônia was called Plan for the Development of the Amazon (PDA I). The program come under the slogan of "*O Homen é a Meta,*" instead of "Amazonia is your best Business" (Bunker, 1985; Hecht, 1985:672). This was a brief interlude in the capitalist penetration of the region. It would bring together "The land without people with the people without land." It aimed at the settlement of small-holding farmers as the solution to social and economic problems (Bunker, 1985:113). The fact that the land was already legitimately owned by Native Brazilians was totally ignored. Indians would have to be "prepared" by FUNAI for their exposure and assimilation to Brazilian society. They would have to be "civilized."

The Transamazon was to be a 4,960 km road connecting the Northeast of Brazil to the Brazilian-Peruvian border (and eventually to the Pacific). Ten kilometers on each side of the highway would be reserved for massive colonization schemes, this area was to be divided into lots of 100 hectares for distribution to incoming settlers. These lots could be purchased for U.S.$700 payable over twenty years with a four-year grace period. For an additional U.S.$100 some settlers were also able to purchase a four-room wooden house for the same financing conditions. As an incentive to attract people to the region, each family was given a monthly salary of Cr$204 (US$30) plus food subsidies, the latter were cut out after six months and the salary increased to Cr$308 (US$50) and extended for eight months. Unlike the tax cuts that were freely given to corporations, these subsidies to small producers were no gift. Settlers were expected to repay what they received within three years (Moran, 1981:79; Smith, 1982:16-17).

Critics called it the road that went from "nowhere to no place," the link between "poverty and misery (Bunker, 1985: 102; Hecht, 1985: 672). As a colonization scheme the Transamazon failed dismally. The several colonization towns constructed along its extent failed to attract the expected number of families, only about 5,700 families had settled in these projects by the end of 1974, less than 10% of the government target. This number had increased to only 8,000 by the end of the decade. Only about 40% of those families originated in the Northeast (Mahar, 1990:63), thus the road failed its purpose of alleviating the land problems of the Northeast. As a matter of fact, in the early 1970's the population of the Northeast was growing by 2.4% a year, 700,000 people each year. While its population grew by at least six million people in the 1970's, the Transamazon only accommodated 23,000 northeasterners, a small 0.3% of the region's population increase. Thus, the road did not act as the safety-valve that had been predicted (Smith, 1982:24).

Several factors may account for the failure of the Transamazon as a colonization scheme. First, there was little *a priori* research on the fertility of the soils along the road. In fact, only about three percent of the soil could be classified as naturally fertile, with most of the area being hilly and subjected to rapid erosion. Erosion necessitated expensive maintenance work for the road and the burning of additional virgin forest to restore soil fertility. Second, malaria, already common in the region, was increased by the favorable *Anopheles* mosquito breeding conditions created by alterations of the forest environment. Third, the upland rice promoted by the government was unsustainable. As soil fertility declined, farmers simply abandoned their fields. Fourth, the colonization projects were far from markets, which put settlers at a disadvantage relative to producers in other regions of the country, and the few roads in the region were impassable during the rainy season. Fifth, distance also determined that the costs of fertilizers and pesticides would be out of the reach of most small producers. Sixth, there were too many bureaucratic obstacles for farmers to obtain credit. Seventh, the settlements were designed by government bureaucrats living far away and who did not understand the customs and preferences of farmers. For example, settlers were not able to raise domestic animals in their plots because properties were unfenced. Also, houses were built far away from the fields, contrary to colonists' preference to live on their lots where they could be close to their fields and domestic animals (Mahar, 1989: 26-27; Smith, 1982:24). It is no coincidence that the only two settlements that flourished along the highway, Repartimento, at Km 160 of the Marabá-Altamira stretch of the Transamazon, and Pacajá at km 212, were free of government (INCRA) control (Smith, 1982:28-29). Finally, there was also a psychological component blocking migration to Amazônia, especially from the Northeast:

> The job of moving any sizeable group to the Amazon is a complex problem.
> Early in the project, the government recognized that the Amazon region was

much feared and mythologized, and that most earlier migrants had gone only under drought duress. The northeasterner in particular had built up an apprehension toward migration to the area. The tales of the rubber gatherers who survived debt slavery in the Amazon and returned to their native land gave the Amazon more than its share of colorful facts and fiction (Moran, 1981:78).

The greatest migration was to pre-existing towns, rather than to the new government projects. In 1970 the town of Marabá had 14,585 residents. By the end of the decade that number increased to 40,000. Before the highway the town of Altamira had 5,734 residents. By 1980 it had increased to an estimated 40,000 (Smith, 1982:28). None of these towns created the expected flood of migrants from the Northeast.

By 1974 all pretensions of Man is the Goal proposed by the PIN were abandoned by the government. Peasants were viewed as lacking the skills to develop the region. Technocrats evaluating the development of the region under PDA I viewed settlers in Amazônia with scorn:

> The indiscriminate migration of these populations, thus, far from making a contribution to the development of Amazonia, creates every year a growing problem of how to absorb in productive employment the labour which manages to settle legally in the area and demands enormous resources in education, rural extension and training. . . . Summing up the question in simple terms, direct and indirect incentives to migration from the north-east merely result in geographical transfer within the country of a problem already established in another region, taking to Amazonia the responsibility of recuperating the so-called surplus population of the north-east . . . (in Branford and Glock, 1985:69).

The government returned to a policy of big projects. This time, however, the projects were even bigger. They became poles of development based not on the needs of landless Brazilian peasants but on the interests of big corporations and the paranoid schemes of the Brazilian military. These projects would center around areas of Amazônia where huge deposits of minerals were discovered and frontier areas which the military deemed to be of importance for national security. The shift to bigger projects did not stop peasants from coming into the region, but they were no longer invited by the government. The migration to parts of Amazônia was made up of people escaping poor conditions in other parts of Brazil or who were, as in the case of gold prospectors into northwestern Amazônia, attracted by the riches offered by the forest. There was little that the government could do to avert this migration, or was willing to do since it brought Brazilians into the region. Thus, the aim of these projects was the immediate extraction of wealth from Amazônia via large corporations; how people were accommodated around them was of secondary importance.

The Carajás Project

The *New York Times* of November 18, 1981 stated the following: "Even for Brazil, where the national leaders like to think big, the project now getting under way at this Amazon site is immense." Projeto Grande Carajás (PGC) occupies an area almost 900,000 square kilometers, 10.6% of the Brazilian national territory, or the size of Britain and France combined. It includes an iron-ore mine, two aluminum plants, and the Tucuruí hydroelectric scheme on the River Tocantins (Hall, 1989:41-42).

Carajás contains the largest high-grade iron-ore reserves in the world, estimated at eighteen billion metric tonnes with an average grade of 66% Fe. It was found virtually by accident when a geologist working for Companhia Meridional de Mineração, a subsidiary of US Steel, stumble upon it when his helicopter landed in the area for refueling. US Steel's request to the Brazilian government for exploration rights was first denied, but a joint venture was later approved, via the creation of Amazônia Mineração SA, AMZA, a conglomerate in which the state-owned Companhia do Vale do Rio Doce (Rio Doce Valley Company) (CVRD) owned 51% of the shares and US Steel 49%. In 1974 AMZA was given exclusive exploration rights over the region (Treece, 1987:12; Hall, 1989:43). In addition to iron ore, Carajás also contains large reserves of bauxite, nickel, cassiterite, copper, and gold (Hall, 1987:533; *The New York Times*, November 18, 1981).

Implementation of the project required an estimated U.S. $62 billion dollars. Foreign investors were eager to participate in joint ventures. In the 1980's the European community provided U.S.$600 million for iron-ore mining, Japan U.S.$400 million, and the World Bank U.S.$350 million (Hall, 1987:535). Also, "Companies in France, Germany, Belgium, Italy and Luxembourg [signed] long-term contracts with CVRD for secure deliveries of iron ore supplies at favorable prices" (Caufield, 1984:20). Companies eagerly invested in aluminum production as well. Two major integrated aluminum projects were devised for Carajás. The first, Aluminar in São Luís in the State of Maranhão, became a U.S.$1.3-trillion-dollar investment by the multinationals Alcoa (USA), 60%, and Billion Metals (a subsidiary of Royal Dutch/Shell), 40%, the largest privately-funded project in Brazil. The other was ALBRAS-ALUNORTE in Barcarena, near Belém, financed by a consortium of 30 Japanese aluminum smelters and the Japanese government in partnership with the Brazilian state-owned Companhia do Vale do Rio Doce (Hall, 1987:533; 1989:57). Dave Treence captures the essence of these investments in the following passage about the EEC's involvement in Carajás:

> In September 1982, the European community granted the first loan to a non-EEC country. The country was Brazil, and the loan was for US$ 600 million, the largest single share of investment in the Carajás Iron Ore Project. This was not some charitable gesture of disinterested concern for Brazil's economic development, with no strings attached, however. It was secured on the understanding that a third of the iron ore from Carajás would go to supply steel works in five European countries (1987:21).

The underlying principle applies throughout. While the project was still in the drawing board, 26 million tons of iron ore had already been sold for delivery to Japan, (West) Germany, France, and Italy (*The New York Times*, November 18, 1981). The primary motivation for the loans was the guarantee of a cheap source of metals which enhanced the lending countries' competitiveness in the international market. To guarantee the flow of these metals a 900-kilometer railway connecting Carajás to the town of São Luís in the state of Maranhão was built in 1985. A year later the new deep-water port of Ponta Madeira in the same city opened. By January 1986 Carajás had produced 13 million tons of iron ore, 92% of which went to Japan, (West) Germany, and Italy (Hall, 1989:48-51). This translated into hard currency that could be used to alleviate Brazil's mounting foreign debt.

In addition to the huge reserves of iron ore in Carajás, Amazônia is believed to contain the world's largest bauxite reserves, 2.2 billion tons. Again, exploration would take place through a partnership between multinationals and Brazilian companies. The Japanese, for example, embarked on a U.S.-$1.3-billion venture through a partnership between Nippon Amazon Aluminum Company (NAAC), a consortium of 33 Japanese firms, and the CRVD (Fearnside, 1990: 202; Hall, 1989:55). For the manufacturing of aluminum a source of electricity would have to be developed.

The Tucuruí dam in the Tocantins river, the largest dam ever built in a tropical rainforest, flooded an area of 2,430 square kilometers. It was put into operation in 1984. The World Bank refused to finance it due to the political and environmental controversies surrounding it, especially the flooding of Indian lands such as the Parakanã Indian reservation and the protest associated with it (Hall 1989, 55; 1985: 535; Fearnside, 1990). The funding for the project was finally obtained through an agreement with the Japanese with the promise of cheap electricity for companies in the complex. Electricity was provided more cheaply to multinational corporations than to Brazilians living in the area, most of whom had no access to electricity:

> One of the most controversial features of the Tucuruí dam is that the power generated does little to improve the lot of those who live in the area: a fact dramatized by the high-tension lines passing over hut after hut lit only by the flickering of kerosene *lamparinhas*(Fearnside, 1990: 203).

"ALBRAS [paid] the lowest tariff in Brazil, obtaining a 15% discount in addition to the already existing concessionary rates for aluminum smelting, and also [enjoyed] a ceiling to its electricity costs of 25% of the world market price of aluminum" (Hall, 1989:55).

Tucuruí had other problems. Only about five percent of the area was cleared, despite recommendations that 85% of the vegetation be removed (Fearnside, 1990: 203). This five percent consisted of species considered valuable, the balance was left to rot. This led to anaerobic decomposition, siltation, weed-clogging of turbines, and acid waters (Hall, 1989:59). The construction of the dam also displaced 35,000 people, who received little or no compensation. As we shall see in Chapter 6, the dam so enraged the Indians in the area that they used its problems as ammunition against other proposed dams.

In addition to development of the mining potential of the Carajás region, the project also called for the development of an agricultural complex that would take advantage of the railroad built for the mining complex. CVRD envisaged Carajás as an export corridor costing U.S.$11 billion dollars. This would include agriculture, ranching, and forestry on huge estates covering 10 million hectares, with a third of the area to be divided into 300 cattle ranches of 10,000 hectares each. In addition, it would also entail large-scale rice cultivation (four million hectares), and that of sugar cane and manioc (2.4 million hectares) (Hall, 1987:535). Also, 1,800 square kilometers along the railroad corridor would be set aside for eucalyptus plantations. The purpose of these plantations was to supply charcoal for the iron-ore smelters, thus limiting environmental damage to the forest. This plan failed because the government had little control of the *carvoeiros*, or charcoal producers, operating within and outside the project area. It is estimated that in order to fuel the smelters 2.6

million hectares of eucalyptus forests would have to be planted (Fearnside, 1988b:100)

Keep in mind that a major rationale for Carajás was that it would help solve Brazil's foreign debt problem. Export earnings would help generate a trade surplus and help service the country's mounting debt (Hall, 1989: 75). This is a vicious cycle: borrowing abroad to finance projects which create debt, all to maintain an addictive relationship with international lenders.

POLONOROESTE

The Northwest Brazil Integrated Development Program, or POLONOROESTE, was established to introduce order to a natural migration flow to the northwestern state of Rondônia, and later to the state of Acre. Hundreds of migrants were arriving in the region, mainly from the South and Northeast, through the recently constructed Cuiabá-Pôrto Velho (BR-364) highway (Mahar, 1989:40). The road was unpaved, however, and thus quite impenetrable during the rainy season. The project was created officially on May of 1981 by Decree no. 86.029. Unlike Amazonian projects administered by SUDAM, POLONOROESTE was under the Superintendência do Desenvolvimento da Região Centro-Oeste (Superintendence for the Development of the Center-West Region) with a budget for the period 1981-1985 of U.S.$ 1.5 billion. It called for the paving of BR-364 and the expansion of the road networks in the area. It also called for the consolidation of colonization schemes already in existence and the development of new ones, regulation of land titles, and support for the economic activities through rural credit and technical assistance. The project counted on the support of the World Bank which provided a third of the total (Mahar, 1982:32; 1989: 34; see also Branford and Glock, 1985:206). By 1981 the bank had approved loans mounting to US$320 million: destined for road construction (US$240 million), agrarian development and environmental protection (US$67 million), and health (US$13 million) (Mahar, 1982: 25, 31-32).

The attempt met with mixed results. The migration flow was too strong to be channeled into colonization schemes by the government. Modernization in agriculture in southern states like Paraná displaced hundreds of families who then moved to Rondônia. The trickle of migrants of the early 1970's increased dramatically by the end of the decade. By the late 1970's the government acknowledged its inability to control the migration and created the "Rapid Settlement" program, which amounted to legitimizing spontaneous settlement. For each family which settled under government supervision another settled on its own. It is estimated that between 1980 and 1985 at least 18,000 new agricultural establishments were created in areas outside government control (Martine, 1990:30).

Brazil also tried to appease international opinion shocked at rates of deforestation caused by peasants clearing the forest by burning it. According to Dennis Mahar (1989:34), "The available data indicate that the actions carried out under POLONOROESTE [did not slow] the pace of deforestation nor appreciably altered traditional patterns of land use." The deforested area of the state increased from 3% in 1980 to an estimated 24% in 1988, to a large extent a result of the land-tenure policy of the government. Deforestation was accepted as evidence of land "improvement," thus qualifying settlers to certain rights and eventually a land title. As the price of land soared, this practice led to increased speculation. Settlers would clear the land and claim it, only to sell it to newcomers. New settlement areas had a turnover of 40% to 55% (Mahar, 1989:36-39).

> Calculations made by the FAO/World Bank Cooperative Program (FAO/CP), which recently reviewed the interim results of POLONOROESTE, show that it is possible for speculators to net the equivalent of US$9,000 if they clear fourteen hectares of forest, plant pasture and subsistence crops for two years, and then sell the rights of possession acquired by doing so. This constitutes a large sum of money in Rondônia, where the average daily farm wage is equivalent to less than US$6.00 (Mahar, 1989:38).

The Calha Norte Project

The Calha Norte project differs from the projects described above because it was devised to address military concerns about the vulnerability of the northern frontier. The project covers a 6,500-km frontier between Brazil and Colombia, Venezuela, Guyana, Suriname, and French Guiana. It includes roughly 14% of the Brazilian national territory, or 24% of legal Amazônia (Hecht and Cockburn, 1989: 118; Allen, 1992: 73). The first attempt to develop the area took place when, in 1973, president Emílio Garrastazu Médici initiated the Northern Perimeter Highway, which was meant to stretch from the Atlantic coast of Amapá to the border with Bolivia in the state of Acre (Pinto, 1988:41). It was to complement the Transamazon highway and provide access to the Northwestern frontier. Construction was abandoned in the late 1970's for financial and technical reasons (Allen, 1992:85), even though "...the contractors earned their money but only managed to complete one-fifth of the first stage of the public work. Nature in the subsequent 10 years of neglect, took upon itself the task of finishing off the partially completed road" (Pinto, 1988:41). But fears of communist infiltration from countries such as Surinam as well as the possible creation of a Yanomami nation (unifying Yanomami peoples on both the Venezuelan and Brazilian sides) reignited the military's desire to develop the region. The project proposal was presented to President José Sarney on June 15, 1985 by a working group which worked under strict secrecy and which consisted primarily of military leaders (Cultural Survival Quarterly, 1988, 39; Treece, 1989:225; Allen, 1992:73; Oliveira Filho, 1990: 156-157).

Under the guise of "national security" the Brazilian people were kept uninformed. The project was classified as "top secret." Congress was not officially notified until October 1987, two years after the project was born. Brazilians vaguely new about its existence because of the denunciations of the Indianist Missionary Council (CIMI), a Catholic missionary organization who exposed the potential harmful consequences for the Indians (Oliveira Filho, 1990:157). Interestingly Brazil was already making a transition to democratic rule. However, despite civilian rule, the government was still largely controlled by the military (see Chapter 5).

Most of the resources, 46% of the total, were allocated to the army for the building and maintenance of barracks, military equipment, purchase of vessels for river transport, and social infrastructure, e.g., education and health. The Ministry of the Navy was allocated 21.4% of the funds for the construction and operation of river patrol boats at the naval base of Val-de-Cans, in Belém do Pará, and a base on the Rio Negro. The Ministry of the Airforce was allocated 10.5% of the funds for the construction and improvement of airports and landing strips, and the maintenance of air service to frontier units. Altogether, the military account for 78.2% of the total budget. Smaller shares were allocated to other government branches and agencies, e.g., 2.1% to the Ministry of Foreign Affairs for demarcation of borders, expansion of consular networks, etc. (Oliveira Filho, 1990: 160-161).

Bruce Albert (1992) sees it as an attempt by the military to redefine Indian policy in Brazil in favor of economic interests, especially mining interests. Even though the project was proposed in ecological and nationalistic language a closer examination reveals an attempt to reduce the size of the proposed Yanomami reservation. It included provisions that classified Indian lands near frontier areas as a special case due to national security (see Cultural Survival Quarterly, 1988:39). The responsibility of integrating the Indians into Brazilian society fell exclusively to FUNAI.

FUNAI's activities in the area were to be kept secret. Missionary organizations dealing with Indians were to be excluded from the area. These were perceived as enemies of Brazil because their assistance to the Indian cause was believed to be against national interest (Oliveira Filho, 1992:161).

The proposal did much to undermine the rights of the Yanomami to their ancestral lands. By overlapping Yanomami Indian lands with "conservation" areas, inter-ministerial directives 160 and 250, calling for the demarcation of Yanomami lands, made the status of the reservation ambiguous and thus subject to future intrusions by commercial interests:

> Directive 160 did not, as FUNAI would have the public believe, give legal protection to Yanomami lands. Quite the contrary, it establishes an ambiguous territorial and administrative arrangement, the whole purpose of which is to appease national and international protests, while at the same time making way for their progressive expropriation (Albert, 1992:40).

Contradicting the mandate of the new Constitution of 1988, which requires approval of the exploration of natural resources on Indian lands by the National Congress, and thus subject to debate, directive 250 states that the activities of non-Indians in Indian lands are subject to the approval of the Instituto Brasileiro do Meio Ambiente (IBAMA) or the Brazilian Institute for the Environment--then the Brazilian Institute of Forestry Development (IBDF)--and FUNAI. This meant that "...the 50 per cent of Yanomami lands that [had] been transformed into National Forests by Directive 250 [could] be opened up *ex officio* to placer mining or to mining companies directly by IBAMA with FUNAI's consent" (Albert, 1992:44). Thus the project served the needs of corporations whose interest in the region appeared threatened. It also gave legitimacy to the invasion of Yanomami land by thousands of small gold prospectors. Mixing of Yanomami land with national forests also gave the impression that the Yanomami reservation was bigger than it was.

Consequences

The consequences of the projects described above involve environmental destruction and human costs. The environmental destruction is identified by the area which has been deforested (see Table 1 in Chapter 1) By January of 1978 an area of 152,200 Km^2 had been cleared as a result of government projects:

Government figures attribute 38 percent of all deforestation in the Brazilian Amazon between 1966 and 1975 to large-scale cattle ranches, 31 percent to agriculture, and 27 percent to highway construction. The government gave fiscal incentives to 90 percent of the ranches, and more than half the agricultural clearings was under a government-sponsored peasant colonization program. That program has now been wound down, in favour of investment in large- scale logging operations, hydroelectric dams, mines, and industrial developments (Caufield, 1991:40).

By April of 1988 the cleared area had increased to 377,500 Km^2, and to 415,200 Km^2 by August of 1990. Some states were more affected by deforestation than others, and the rates varied over time, being affected by government policy and programs. Through the 1970's and 1980's the states most affected were those crossed by the Belém-Brasília highway: Tocantins (then part of the state of Goiás), Maranhão, and Pará. In 1978 these states together accounted for 123,500 Km^2 or 81% of all the deforestation taking place in Amazônia. The other major focus of deforestation lay west of Brasília, in the state of Mato Grosso. In 1978 Mato Grosso alone accounted for 13.14% (20,000 km^2) of the total cleared area. By 1988, it accounted for 18.9% or 71,500km^2. Much of this deforestation can be attributed to the economic activities generated by the Brasília-Cuiabá highway. In the 1980's deforestation advanced westward to the states of Rondônia and Acre and to the POLONOROESTE region. By April 1988 Rondônia and Acre together contributed 10.3% of the total deforestation of Legal Amazônia, 38,900 km^2, compared to a total of 4.4%, or 6,700 km^2, in January of 1978.

In the 1990's the patterns changed slightly, with deforestation spreading still further west. The combined cleared areas of the states of Maranhão, Pará, and Tocantis declined from 81% of the total in January 1978 to 62% in 1996. On the other hand the area for the state of Mato Grosso increased from 13.14% (20,000 km^2) to 23% (119,141 km^2) in the same period. For Acre and Rondônia the increase of 62,390 km^2 was modest, from 10.3% of the total by January 1978 to 12.1% by August 1996. The states of Amapá, Amazonas and Roraima, together, accounted for only 1.3% (2,000 km^2) of the total cleared area of legal Amazônia in January of 1978. By August of 1996 they accounted for 34,577 km^2 or 6.7% of the total (27,434 km^2 or 5.3% in the state of Amazonas alone).

According to INPE (1998) the main causes of deforestation in the region have been the conversion of forests to pasture and agriculture, especially the production of grains. The biggest expansion occurred in the states of Mato Grosso, Pará, Rondônia, and Tocantis. INPE also attributes the deforestation to the food needs of the 20 million inhabitants of the region, done primarily by small producers. These small producers fertilize the soil by burning the biomass contained in the forest. INPE also lists the price of land and logging as causes. It argues that declining land prices discourage investments in intensive agriculture and sustainable logging practices. Peasants simply clear new plots when resources dwindle. Because of invasion of idle properties by landless peasants, land owners clear their land of forest in order to establish more solid claims to their properties. In terms of logging, INPE points out that 90% of the wood consumed in Brazil comes from Amazônia.

The Human Consequences

These projects have had severe consequences for the people of Amazônia. The most dramatic impact has been felt by the Indigenous populations of the region. Their lands have been invaded by non-Indians who brought with them disease and violence. The area occupied by Projeto Carajás, for example, contains an estimated 40 or so Indian communities inhabited by a total of 13,000 tribal people. Their territories are being coveted by big farmers and developers as well as by peasants (Treece, 1987:6).

> The invasion and destruction of forests which these projects inevitably bring with them does not only signify a loss of territory and security for the Indians. It also triggers a whole series of interrelated social, health and cultural problems. The Parakanã communities have been subjected to eleven moves in all, first in order to settle them after initial contact and then to relocate them after the Tucuruí reservoir flooded their lands. As a result they have been forced to abandon their traditionally nomadic lifestyle, leaving them dependent on the official aid provided by FUNAI and exposing them to diseases which did not previously affect them, such as malaria. A quarter of the Indians in the Marudjewara village were killed by the disease within six months in 1983, while eight years of contact reduced Guajá of the Carú Indian Post from 120 to 29 in 1980. Settlement has also made the Xikrin, Asurini and Saruí vulnerable to water-borne disease like river-blindness and schistosomiasis, and to intestinal infections encouraged by the accumulation of waste (Treece, 1987:56).

This account is representative of Indian contact in Amazônia (see for example O'Connor, 1997; and Branford and Glock, 1985). The Indians are the biggest losers.

As we shall see in Chapter 6, people involved in extractive activities, such as rubber tappers and brazilnut extractors, have also had their livelihoods threatened. As big landowners have moved into the region they have attempted to take over lands traditionally used by these people. The result has been escalating violence in states such as Acre where extraction is an important economic activity. Large landowners compete with small migrant peasants for land, often engulfing the small properties. Peasants have also invaded land that belongs to absentee landlords through the support of the *Sem Terra* (Landless) movement of Brazil. The result has been severe violence in states such as Pará where hundreds of people have been killed in land disputes. In a very publicized case that took place in April, 1996, Brazilian police "lost control" and killed at least 23 landless workers while injuring 50 others in a clash on a remote Amazon highway. The incident occurred when about 2,500 landless peasants demanded land near Eldorado dos Carajás, a small town 693 kilometers south of Belém in the state of Pará (Nando.Net, 1996).

The Ecopolitics of Project Building

What is striking about the projects discussed above is that until the 1980's they were devised and implemented by the government without any public discussion as to their significance. During the dictatorship, the assumption was that such projects were in the interest of Brazil because they were de facto incorporating Amazônia into the national territory, thus reducing the risk of foreign invasion. Those opposed to these development schemes were few and easily stopped because individuals had little political freedom and the media was censored. Television was to a large extent monopolized by the powerful Globo network, which favored the developmentalist philosophy of the government. Globo shows such as "Amaral Neto o Reporter" ("Amaral Neto the Reporter")[1] made a sensation out of development projects, arousing a sense of nationalism and pride in the future of the country. Newspaper articles written questioning the validity of such projects went unnoticed by the bulk of the population. This was not the age of Brazilian environmentalism. The few organizations that existed were apolitical and concerned predominantly with urban issues. Pollution in urban environments affected more people and was a problem more easily identified by most Brazilians. Cities such as Cubatão in the state of São Paulo were so polluted that they inspired a popular soap opera in the 1970's, "Cidade de Pedra" (City of Stone). Even this soap opera failed to motivate the population to join the environmentalist movement.[2]

The idea of preserving nature for the sake of nature remained an alien concept.

Until the 1980's the ecopolitics of the world-system favored development projects over preservation of tropical ecosystems. The cries against the destruction of tropical rainforests were largely unheard by the public at large. There was little interest in the plight of rainforests in the developed world. Organizations such as the World Bank favored large development projects and concerned themselves little with impacts on the environment and indigenous peoples. Internationally the Brazilian government had allies, not enemies, in the process of destruction.

This situation changed drastically in the 1980's with the advent of democracy in Brazil, and heightened international public concern over the deteriorating state of the global environment. While the devastating environmental impact of the early projects was hardly noticed, either in Brazil or the world, those initiated in the 1980's gained much notoriety as scholars and more people became aware of the importance of tropical rainforests to the climatic stability of the planet. We now need to address the changes taking place both within Brazil, the transition to democracy-- and within the world-system-- the transition to (a rhetoric of) sustainable development.

Notes

[1] This 1970's series, narrated by journalist Amaral Neto, showed different features of Brazilian life, often from a very nationalistic angle. One of its most controversial segments was about the whaling industry of Brazil, showing the actual killing of whales along the Brazilian coast.

[2] One has to understand the mass appeal that soap operas have in Brazil. They are shown at prime time and are viewed by millions of viewers.

Chapter 4

Changes in Global Ecopolitics

As argued in Chapter 1, capitalism involves a whole mind-set geared towards the production and consumption of goods and the profits that this process entails. And while capitalism is also an economic system based on the rational pursuit of profit, often this very rationality can also lead to irrational outcomes. What is rational at one time can become the irrationality of the next. For example, because fossil fuel is cheaper than alternative fuels for automobiles it has been, from the point of view of profits, the rational fuel to be sold in the market. From an environmental point of view, however, the existence of alternative sources of energy in the age of the greenhouse effect renders its use irrational. Why then do we continue to use it?

In any system, those who benefit from the status quo are not easily persuaded to change or adopt alternative behaviors or strategies. Governments, assuming the rationality and benefits of capitalism, are reluctant to switch to policies that their corporations find dubious, that might threaten the economic stability of their countries or, in the case of developing countries, their entire development prospectus.

On an individual level people are often reluctant to give up consumer goods for environmental alternatives which they may perceive as less efficient and perhaps less glamorous. Riding bicycles or taking public transport rather than driving cars, for example. It is also hard to convince people in the developing world to leave existing ecosystems such as tropical rainforests intact when the land could provide them and their children with a better life. Without leadership from their governments, without alternatives, they will continue to see these lands as free for the taking. People in these countries are also being socialized into a culture of consumerism through the power of advertising. To consume, or the idea of one day becoming a consumer, has become a sign of prosperity for them as well. The truth is that people like the material things that capitalism is capable of providing; even those who cannot obtain these goods, due to poverty, dream of the possibility of one day having them. In a world in which success is increasingly measured by material possessions behavioral and policy changes do not come easily. Even when people feel in their hearts that change is needed, their governments are too slow to provide the appropriate infrastructure and technology for changes to occur, e.g., providing the means for people to recycle. Governments often act conservatively because they are influenced by the interests of systemic agents of destruction such as multinational corporations and elites who profit from the status quo. Major ecological problems beginning in the mid-1980's have forced governments to take a serious look at the environmental situation despite the pressure from corporate elite interests.

In the mid-1980's environmental disasters such as the Exxon Valdez incident in Alaska, and Bhopal in India, made many people stop and think about the environmental consequences of economic activities. These events were widely publicized by the media, which showed powerful images of people dying in Bhopal and birds, mammals, and beaches covered with oil in Alaska. But, these were specific events afflicting particular parts of the world, their consequences were not felt everywhere, not to the extent that most people could see or feel them. But two other environmental problems were more extensive: the hole in the ozone layer and the greenhouse effect. Unlike the Exxon Valdez and Bhopal incidents these global problems are less easily ignored. When the bodies in Bhopal disappeared from view and the beaches were no longer covered with visible oil in Alaska media attention declined. These are disasters better remembered by the local peoples and environmentalists. On the other hand, the ozone layer hole and the greenhouse effect are perceived as truly international problems. It is becoming increasingly understood that they have potentially devastating consequences for everyone on the planet. They cannot be "cleaned" like the beaches of Alaska, or cremated like the bodies in Bophal. They are problems that alarm governments because of their enormous potential economic and social consequences, e.g., floods, droughts, skin cancer, rising sea levels, etc. More than any other environmental concerns these two have had a major impact on global ecopolitics in recent years. They have become symbols of the precariousness of the environment. They have forced us to think about contributing factors such as deforestation and pollution. Increased awareness about the state of the environment now appears to be a phenomenon of worldwide dimensions.

Increased Awareness

By the late 1980's the media coverage about the declining state of the environment had increased public awareness about the severity of the situation. Robert Mitchell (1990:88-87) reports a significant increase in public concern about the state of the environment from the 1970's to the 1980's. The percentage of people rating the item "Improving and Protecting the Environment" as one of several items under "We are Spending Too Little to Solve National Problems," increased from 45% in 1973 to 48% in 1983. By 1988, 62% agreed that we are not spending enough to solve environmental problems. Mitchell attributes much of the increased awareness to the media's growing interest in environmental stories, a new resource which environmentalists used to their advantage:

> Once the media's growing appetite for environmental stories was stimulated, environmentalists did their best to feed it. In the early spring of 1989 a report issued by the Natural Resources Defense Council about the potential risk posed to children by pesticides, especially Alar, created a big media splash and made the cover of *Time*.... In late 1989 the Wilderness Society and the Sierra Club, after years of effort and the help of the endangered northern spotted owl, succeeded in making logging of old-growth or virgin timber in the national forests of Washington and Oregon a newsworthy topic that found its way to the cover of *U.S. News and World Report* (Mitchell, 1990:89).

Another indicator of this increased concern was the sharp rise in membership in environmental organizations in the U.S. in the 1980's, especially after 1983. For example, Helem Ingram and Dean Mann (1989:139-140) documented increased membership in organizations such as the National Wildlife Federation, the Sierra Club, and the Wilderness Society in this period. Membership in the Sierra Club increased from 35,000 in the 1960's to 180,000 in 1980. By 1983 its membership had climbed to 346,000. Membership in the National Adubon Society was up 70% in the same period.

Increased environmental awareness was not restricted to developed countries. The results of a 1992 "Health of the Planet" (HOP) survey by the

George H. Gullup International Institute show that people throughout the world were aware of the major environmental problems afflicting the globe. For example, in North America (the United States and Canada) 63% of those surveyed rated the "Loss of Ozone" as a "very serious" problem, 52.5% gave Global Warming the same rating-- 67% for "Loss of Rain Forest," 53.5% for "Loss of Species," 55.5% for "Contaminated Soil," 74% for "Water Pollution," and 60.5% for "Air Pollution." The results for Latin America (Brazil, Chile, Mexico, and Uruguay) show a similar high concern about the seriousness of these problems. Of those interviewed 76.75% rated "Loss of Ozone" as a "very serious" problem, 76.75% "Global Warming," 62.25% "Loss of Rain Forest," 77% "Loss of Species," 75.5% "Contaminated Soil," 66.25% "Water Pollution," 75.25%; and "Air Pollution," 74.5% (Dunlap, Gullup, and Gullup, 1992: Table 6).[1] Even though the results show regional variations and despite the complications associated with conducting surveys on such an international scale, it is apparent that people throughout the world, not only in the industrialized countries, are becoming aware of major environmental issues. The HOP results compelled Riley E. Dunlap and Angela G. Merting (1995) to conclude:

> Our results not only challenge "lay" wisdom, but also conventional social science analyses of environmentalism. The idea that environmental quality is a luxury affordable only by those who have enough economic security to pursue quality-of-life goals is inconsistent with the observed correlations, as well as with the overall high levels of environmental concern found in residents of the low-income nations in the HOP. Our results are thus compatible with growing evidence of strong, grass-roots' environmentalism in the Third World..., and suggest that such activism likely reflects broad-based concern for environmental quality in these nations rather than the anomalous acts of small minorities (p.134).

They are critical of the prevailing view in the social sciences that environmental awareness is a monopoly of the rich countries and their post-material culture:

> It appears that conventional social science perspective on global environmentalism are in need of revision. Theories on the emergence of postmaterialist values and new social movements appear useful for explaining environmentalism within most wealthy, industrialized nations..., but seem unable to account for the emergence of grass-roots environmentalism and widespread concern for environmental quality in poorer nations. In part this may stem from the fact that environmental

degradation is increasingly seen, especially in poor nations, not as a postmaterialist quality-of-life issue but as a basic threat to human survival...(p. 135).

In developing countries, then, the "threat" has compelled people to action, especially for people whose survival is already being threatened, such as Native peoples and rubber tappers in Brazil. The growing number of environmental NGO's in the developing countries attest to this awareness (see Princen and Finger, 1995: 2).

NGO's Proliferate and Politicize

Motivated by the severity of the global environmental situation and taking advantage of the opportunity created by the intense media coverage of environmental issues from the mid-1980's to the early 1990's, existing NGO's began to participate more forcibly in global politics, while new ones were created to address specific problems. The attention they received from the media provided the opportunity to mobilize resources. Northern NGO's amassed public support which translated itself into financial support. With a bigger financial base they helped Southern NGO's, often their own subsidiaries:

> Mounting public support in the rich countries of the North has given their NGOs a much stronger financial base. Northern NGOs collectively now transfer to the South more than the World Bank group does. NGOs have frequently demonstrated their ability to help those most in need who have been missed by official aid programs. And most important they have pioneered new approaches and challenged development orthodoxy. In particular many Southern NGOs have successfully challenged socially environmentally damaging programs pursued by their own governments (Clark, 1991:3).

It is no coincidence that only about 30% of all development NGO's in the developing world are more than 30 years old, and only 50% are more than 10 years old (Livernash, 1992:14). Not only did NGO's grow in size and in numbers throughout the world in the 1980's, they also "...changed substantially, moving away from an exclusive focus on small-scale projects bounded in space and time, towards increasing involvement in the broader process of development... (Edwards, 1994: 117). On a global level, NGO's became collectively the most active and powerful anti-systemic force. This became possible when they joined their political efforts, forming "umbrella organizations" and "networks." It is important to note that "Almost all environmental NGOs and most NGO networks and umbrella groups started in the 1980s..." (Livernash, 1992:14). Their strength allowed them to gain attention for their challenge of the prevailing view of development as usual--against the irrationalities of capitalism. International organizations and governments were forced to take notice of the harmful impact of development on the environment. And they made their target the most powerful development agency: the World Bank.

Critics of the Bank accused it of helping the rich and multinationals rather than the majority of the people of the third world, the poor. They were critical of the Bank's "trickle-down" policies. Critics argued that not only did the poor seldom benefit from such investments, but that these projects too often made their lives worse, e.g., dislocating population to make way for dams (see Helmore, 1988; Holden, 1987). In particular, the Bank was subjected to mounting criticism for disregarding the environmental impact its projects had on native peoples. One former consultant stated that "One cannot help but feel that the World Bank is much more concerned with images than with the welfare of the native minorities [of Brazil]" (Price, 1985:75).

Despite its powerful position in the global system the Bank had an important weakness: it is heavily subsidized by the Western countries, particularly the United States. Because these countries provide the Bank's financial support, they also hold the majority of decision-making votes. This control by the developed countries has made the Bank a vehicle for keeping third-world countries within the western sphere of influence (Barbosa, 1993a, 1993b; Kolk, 1996:186). Thus, NGO's attempted to influence Bank decisions by putting pressure on the supporting governments, especially the U.S. government. The Bank was now embroiled in a global dialectical process in which it could no longer ignore the environmentalists without paying the price of protests and political pressure. The Bank responded positively when in 1987 its president, Barber Conable, announced a sweeping restructuring of the organization. Despite an elimination of 500 positions overall, 40 new environmental positions were created, expanding the total number to 60 (Hopper, 1988:769). The environmental staff included "...such luminaries as maverick economist Herman Daly...; anthropologist Shelton Davis, a long-time activist on behalf of South American Indians; and Peruvian ecologist Marc Dourojeanni, who is credited with founding the conservation movement in Peru" (Holden, 1988:1610). *Science* magazine called this restructuring the "greening" of the World Bank (Holden, 1988:1610). However, the so-called "greening" of the World Bank did not lessen pressure from environmentalist organizations, nor did it cancel previously granted environmentally unsound loans.

In the United States, on October 4, 1989, largely due to pressure from environmentalist organizations, the U.S. Congress conducted a hearing entitled "Environmental Impact of World Bank Lending" in which the activities of the Bank were severely criticized (U.S. Government, 1989). A key NGO strategy to influence politicians and other decision makers was to bring native peoples and grassroots leaders to the United States and Europe to tell their own stories. Clearly these alliances were priceless for both NGO's and the people of the forest.

Action as a Result of Fear and Pressure

A critical feature of ecopolitics is that it is slow to change, even in the face of scientific evidence calling for change. Governments, corporations, organizations, and average citizens do not easily adopt new environmental policies, even despite a consensus suggesting difficult times ahead. These actors often take "unscientific" self-interest positions in matters regarding the environment (see Tarlock, 1992: 63). The fact that scientists cannot agree on the extent of some problems often provides actors with the excuse to maintain the system as is. We have known for years now that the use of gasoline-engine automobiles contributes disproportionately to pollution and the greenhouse effect. Yet, in countries such as the United States, there is as yet little effort to switch to more environmentally sound forms of transport. Caroline Thomas (1992:224) tells us, for example, that in 1985 lack of conclusive scientific evidence connecting CFC emissions to the ozone problem gave the lobby for large companies the upper-hand in pressuring government to take a moderate stand in the Vienna Convention for the Protection of the Ozone Layer. Similar excuses have been presented by industry regarding the greenhouse effect. The reluctance by the United States in Kyoto, Japan in 1997 to sign a treaty on the reduction of carbon dioxide emissions largely reflects the lobbying of corporations to reduce the economic impact of the treaty on themselves.[2]

Systemic agents almost always attempt to maintain the status quo because their profits are often connected to environmental destruction. But environmental "manifestations" (Barbosa, 1990) will force people to action. These "manifestations" are the more visible and eminent environmental problems that create concern and fear among people, such as droughts, El Ninos, hurricanes, tornadoes, and floods. They make us stop and think about what is happening around us. When, for example, the media reports that the year 1998 was the hottest year on record and connects this to global warming, it concerns people about the future; it makes governments move towards addressing these problems. However, even "manifestations" are no guarantee that an environmentally rational course of action will be pursued. People may simply bury their heads in sand. Unless they are affected directly via the loss of property or lives, human beings have very short memories.

Attempts to address global environmental problems are not new. The deteriorating state of the environment led the United Nations to organize the United Nations Conference on the Human Environment, in Stockholm in June of 1972. This Conference was punctuated by a major North-South political divide on issues such as population, cleanup costs, and sovereignty. These disagreements reduced the general impact of the Conference. Its major accomplishment was the creation of the United Nations Environmental Program (UNEP) and the Environmental Liaison Center (ELC). UNEP has played a vital role in facilitating discussions on a broad range of international environmental problems between governments, U.N. agencies, and NGO's. The ECL is a coalition of environmental NGO's. It provides a mechanism for inter-NGO communication (Thomas, 1992:24-25). The Conference involved the participation of more than 250 NGO's, the highest number at any Conference until UNCED in Rio de Janeiro (Willetts, 1996: 69-70).

The lack of an imminent global environmental threat, or a problem perceived as affecting everyone on the planet at the same time, meant that after Stockholm the state of the global environment took backstage in international politics until the latter part of the 1970's when the emergence of an urgent and truly global problem forced governments to the negotiation table once again.

In 1974 an article by scientists M. J. Molina and F.S. Rowlandin appeared in *Nature* theorizing about the connection between ozone depletion and CFC's. Their findings were alarming because these gases were until then perceived as harmless (Thomas, 1992:201). As a result two conferences were held in 1977 and in 1978 in Washington and Munich, respectively. These conferences were characterized by a reluctance by many developed countries to reduce the use of CFC's. These countries, however, differed in their respective positions. Canada, the United States, and Scandinavian countries, already had reduced their emissions considerably as a result of a ban on CFC usage in aerosols in 1974. Most European countries took a more conservative position on reduction: "The Europeans, who had made very little effort to reduce their emissions, used the argument of scientific uncertainty to justify their position" (Thomas, 1992:221). Despite this reluctance, however, UNEP pressed ahead with efforts to create a protocol for the use of CFC's. These efforts led to the signing of the Vienna Convention for the Protection of the Ozone Layer on March 22, 1985. The Convention fell short of establishing specific reduction targets and obligations for the signatories. The lack of specific targets and obligations gave leverage to industries lobbying governments to take a moderate stand on the issue.

The announcement of the thinning of the ozone layer, "the ozone layer hole," in 1985 substantially altered the nature of the negotiations. Most people were ready for regulations on the use of CFC's. After extensive negotiations the Montreal Protocol was signed on September 16, 1987. It called for a 50% cut in the use of five major CFC's by 1999, using 1986 as a baseline. In addition, the use of halons would be frozen by 1992 at the level of 1986. The treaty was amended in June 1990 in London. The amendments accelerated the phaseout schedule for CFCs and halons and added methyl chloroform and carbon tetrachloride to the list of chemicals to be eliminated (Thomas, 1992:224-226; Stern, Young, and Druckman, 1992: 116-120; Switzer, 1994:276-279). The urgency of the problem facilitated the development of this international agreement:

> Analysts have identified four important factors [leading to the agreement]: an evolving scientific consensus; a high degree of public anxiety in developed countries about the risks associated with the continued use of CFCs, due in large measure to an association with skin cancer; the exercise of political muscle by the United States, and the availability of commercial substitutes for CFCs. . . . Another important influence . . . may have been the efforts of the scientific community, which has been influential in drawing attention to other environmental problems (Stern, Young, and Druckman, 1992:119).

Other environmental "manifestations" would force countries to begin to address the the greenhouse effect. A hot and dry summer in 1988 in the United States focused the debate about the state of the environment. There were no assurances that the hot summer was due to global warming but this did not prevent the media from making that connection. Thomas A. Sanction (1989) writing for the 1989 *Time* "Planet-of-the-Year" special edition, for example, stated the following:

> This year the earth spoke, like God warning Noah of the deluge. Its message was loud and clear, and suddenly people began to listen, to ponder what portents the message held. In the U.S., a three-month drought baked the soil

from California to Georgia, reducing the country's grain harvest by 31% and killings thousands of head of livestock. A stubborn seven-week heat wave drove temperatures above 100^0F across much of the country, raising fears that the dreaded "greenhouse effect"--global warming as a result of the build up of carbon dioxide and other gases in the atmosphere--might already be under way (p.26).

Michael D. Lemonick (1989) in the same issue suggests a possible scenario resulting from a common fear in the possibility of the greenhouse effect and droughts:

Some forecasters have suggested that the impact of global warming will not be uniformly bad around the world. After all, Canada would not complain if the productive corn-growing lands of the U.S. Midwest shifted north across the border... (p. 37).

Lemonick's article also tells us about the impact that the thermometer had in changing the attitudes of politicians, making them take the threat of global warming seriously:

When Colorado Senator Timothy Wirth held congressional hearings on the greenhouse effect in the fall of 1987, the topic generated no heat at all. "We had a very, very distinguished panel," Wirth recalled at the **TIME** Environmental Conference, "and who was in the cavernous hearing room? Six or seven people, and two or three of them were lost tourists." So Wirth decided to schedule another hearing in the summer, hoping hot weather would make people pay attention to the greenhouse issue. Sure enough, the hearing convened last June 23, the thermometer read 99^0 F, a Washington record for that day. The room was packed when James Hansen, head of NASA's Goddard Institute for Space Studies, turned global warming into front-page news at last (p.36).

The possibility of economic loss in face of natural phenomena such as droughts compelled international leaders to listen to scientists and environmentalists and to question policies of the World Bank and other international organizations. Above all, it led to UNCED.

By 1987 the World Commission on Environment and Development's document, *Our Common Future*, had put forth the concept of sustainable development as we understand it today. The same document also called for a second U.N. Environmental Conference on the state of the environment. In December 1989 a formal decision was made to convene UNCED in Rio de Janeiro, Brazil. The resolution also established a Preparatory

Committee (PreCom), which would include all UN members. It was decided that the committee would meet four times before UNCED and that it would be open to the participation of observers in accordance with the established practices of the General Assembly (Willetts, 1996:70; Union of International Organizations, 1995). NGO's would play a substantial role in the conference that would provide us with the environmental guidelines for the 21st century.

UNCED: Further Steps Towards Paradigm Transition

The importance of UNCED is that it created a series of treaties or documents that will guide environmental politics into the 21st century. The main documents signed during the conference were: 1) Convention on Biological Diversity; 2) Forest Principles; 3) Framework Convention on Climate Change; 3) The Rio Declaration on Environment and Development; 4) and Agenda 21 (see Grubb et al., 1993). All of them were controversial, not only because of divergent opinions between the North and the South, but within each camp as well. The United States, for example, was in opposition to both developing and developed countries. It was reluctant to sign the treaties on biodiversity and climate change because the former called for industry to compensate developing countries for the use of forest products, e.g., plants for pharmaceuticals, and the latter put timetables on the reduction of carbon dioxide. The assumption was that both treaties would harm U.S. industry, while the latter treaty would also lead to unemployment. Despite these controversies the documents were signed by the majority of countries attending UNCED.

The Rio Declaration and Agenda 21 provided us with further steps towards a model of "sustainable development;" an "official" endorsement of the model itself. The Rio Declaration contains 27 general statements about sustainable development including "The right to development must be fulfilled so as to equitably meet the developmental and environmental needs of present and future generations" and "To achieve sustainable development and a higher quality of life for all people, States should reduce and eliminate unsustainable patterns of production and consumption and promote appropriate demographic policies" (in Thompson, 1993: 87-89). Agenda 21 is a more specific action program:

> Agenda 21 is intended to set out an international programme of action for achieving sustainable development in the 21st Century. It seeks to be comprehensive in its scope, and to make recommendations on the measures which would be taken to integrate the environment and development

concerns. To this end, it provides a broad view of issues pertaining to sustainable development, including statements on the basis of action, objectives, recommended activities, and means of implementation (Koch and Grubb, 1993:97).

Together, these documents have provided the basis for further discussion on the state of the environment. For example, when countries met in New York in June of 1997 to evaluate the accomplishments made after Rio in the "Rio Plus Five Conference," the treaties signed during UNCED served as guides for evaluation of national progress. Participating countries were asked by the U.N. to produce a document, "National Implementation of Agenda 21," for the meeting (see U.N., 1997). With these documents as reference critics recognized that the implementation of Agenda 21 has been slow. When countries met in Kyoto, Japan in December of the same year to discuss climatic change, they were also attempting to improve the Convention on Climate Change signed in Rio.

Despite the severe global environmental problems at the time, UNCED was not an easy conference to organize. There were serious divisions between countries. As in Stockholm in 1972, these divisions were especially severe between the North and the South, even though other disagreements existed. The fact is that countries were reacting according to their economic interests. For example, the United States under the presidency of George Bush was reluctant to endorse treaties on carbon dioxide reduction; arguing that implementation would lead to loss of jobs. Countries such as Malaysia and Brazil resisted negotiations leading to restrictions on the clearing of tropical rainforests. Many governments of developing countries were suspicious of the attention being given to the environment. They viewed the conference as another attempt by the developed countries to limit their economic growth (Willets, 1996:72).

One of the key disagreements from the beginning was about participation of NGO's in the preparatory stages of the conference. In 1990 the PreCom held an organizational meeting in which ". . . eleven of the well-established global INGOs put their toe in the door and attended the section . . ." (Willets, 1996:71). Since the role that these organizations were to play was not clearly defined; the committee failed to establish guidelines for their participation in the conference. While the key organizer of UNCED, the Canadian Maurice Strong, was a leading supporter of NGO participation, some developing countries were antagonistic to the participation of NGO's. They viewed these organizations as hostile to their development interests. The major issue confronting the PreCom committee was one of access. How many—and which-- NGO's should have access to the preparatory process? It was decided that all environmental organizations with ties to ECOSOC would be allowed to attend the meetings, as well as NGO's with "special competence" or expertise. In an "unofficial" meeting on 10 August 1990, it was decided that ECOSOC NGO's would have an opportunity to speak in meetings while other NGO's would have to ask to speak, what is known as Decision 1/1. However, this Decision also stated that NGO's would have no negotiating role and no right to issue statements as "official" conference documents. They, however, could distribute written documents at their own expense (Willets, 1996:75).

The restrictions imposed by Decision 1/1 did not inhibit the attendance of NGO's at PreCom II in March-April 1991. Their numbers increased from 41 to 167, though most organizations did not have consultative status. The PreCom II committee also decided to facilitate the process of accreditation for these organizations. The result was that "From then onwards, the NGOS did have an open door and NGO activity blossomed" (Willets, 1996:75; see also Union of International Organizations, 1995).

In addition to participating in the organizational process of UNCED, NGO's maximized their political leverage by organizing a parallel conference, the "Global Forum," during UNCED.[3] By March of 1992, two months prior to UNCED, forum organizers had received commitments to participate from 5,436 representatives for 1,845 NGO's. Of these 1,845 organizations about 41% were based in Latin America, 22% in North America, and 20% in Europe (Livernash, 1992:16). These organizations represented a very wide spectrum of interests, from strictly environmentalist organizations to those only tangentially environmentalist, e.g., AIDS prevention. Despite the often chaotic climate of this forum, it provided NGO's throughout the world with an opportunity to organize in one place and draft documents for the future. Their presence at this event gave NGO's increased political leverage via media exposure. The forum often reached a "*carnavalesco*" atmosphere even though most events were serious. The less formal atmosphere of these forums should not minimize their importance:

> So important have the UN's Global Conferences and Summits become in expanding the NGO role in world affairs that they require closer examination. The pattern at UNCED--of admitting NGOs to a participatory role without going through the formal process of acquiring consultative status--continued for the major conferences thereafter. Each of them-- nutrition (1992), human rights (1993), population (1994), social development (1995), women (1995) and habitat 1996)--has been, or will be,

accompanied by a large and vocal NGO presence, usually organized in a simultaneous all-embracing NGO Forum. But if the overall forum sometimes has appeared to be a disjointed ten-ring circus, many of the special interest groups within it--women's organizations are a striking example--have organized themselves for many months prior to the conference and have come prepared with the platform for which they are committed to. lobby intensively (Union of International Organizations, 1995).

Another "non-official" event of importance in Rio was the World Conference of Indigenous People. This conference was held two weeks prior to UNCED in an Indian village, the Karioca Village, constructed especially for the event in the outskirts of the city. It was the largest gathering of indigenous people in history. There were representatives from many groups: "There were Sarawaks from Malaysia, Hill people from Pakistan, Apaches and Onondaga people from North America, Sumi people from the Nordic countries, aborigines from Australia, Maoris from New Zealand, and dozens of other 'forgotten peoples,' as they have been called in the popular press, including a wide spectrum of indigenous societies from Brazil"(O'Connor, 1997: 298). This event received a great deal of media attention putting the plight of Indigenous people throughout the world in the spotlight:

> But as with any of the large conferences I have attended in recent years, there are early warning signs of the problems to come. For every Indigenous leader here there seem to be at least two or three photographers colliding, cursing, tipping, and shoving, all of them trying to get the best angle, the most intimate exposure. The scene is similar to Altamira, but more extreme: there is more hysteria, more tension. Given the current climate of interest in tropical forests, images of Indigenous people are hot-ticket items. These photographers seem to be particularly interested in Brazilian Indians whose elaborate ritual outfits have attracted the largest hordes of image gatherers (O'Connor, 1997:299).

All of a sudden people whose plights had long gone unnoticed were the center of media attention in a political forum of global proportions. At least for the moment the world took notice of them, though the native peoples of the world were not allowed much time to express their grievances in the "official" conference. Yanomami Indian leader Davi Kopenawa, like anyone else, had about 10 minutes to express the frustrations. In the beginning of his speech he complained about the amount of time allocated to the Indians despite their centuries-old persecution.[4]

If one takes into consideration how much time the Native peoples received to speak in the "official" conference, it is obvious that world leaders were not the most concerned about their opinion, the opinion of people most affected by development projects. It appears that their participation in the official conference was mostly a symbolic gesture, their views on environmental issues not considered important.

The official conference, or UNCED, took place in another part of town miles away from the Global Forum. As already stated, the Conference was punctuated by major disagreements between the North and the South. In addition, there was lack of leadership by the United States. The Bush administration treated the conference with a high degree of contempt. There were rumors that President Bush would not attend, a threat he used in order to obtain concessions in the language of treaties. His arrogant attitude made the United States the villain of the conference. During his official speech to UNCED he stated, "We did not come here to apologize" (*Jornal do Brasil*, June 13, 1992).[5]

The Bush administration viewed UNCED as a threat to American corporations and jobs. It refused, for example, to sign a treaty on carbon dioxide emissions reduction claiming that its timetables would force American corporations to cut their work force, thus creating high unemployment. It also refused to sign the Convention on Biological Diversity on the grounds that it was bad for U.S. corporations. (This treaty was later signed by the Clinton administration.) This self-interested stance by the United States may have made it easier for smaller countries such as Malaysia to argue along similar lines, forcibly challenging the language of the Forest Principles. The U.S. worked not in favor of strong language but weaker language, setting an environmentally-unsound precedent. It frequently clashed with its European counterparts who were more favorable to stiffer language and timetables.

After the media fanfare, prior to and during UNCED, the number of news pieces about the environment declined. News about localized conflicts in places like the Amazon rainforest once again were largely ignored. However, the environment itself would not be ignored. Major climatic problems have continued to remind us that global warming has not gone away. The environment continued to manifest itself in the form of forest fires, El Ninos, droughts, and floods, forcing environmental leaders to meet again in Kyoto to discuss climatic change, where old differences resurfaced. Major systemic agents of destruction once again attempted to influence the course of decisions. American corporations bought commercial time on CNN network in an attempt to influence public opinion on carbon dioxide emissions. Without acknowledging that the United States is the biggest emitter of carbon dioxide and that it has achieved its high standard of living by polluting the atmosphere, these commercials argued that all countries should be required to reduce emissions, whatever their stage of development. Capitalist interests were again trying to maintain economic benefits, a stance that will become increasingly difficult to justify as the state of the environment declines. As environmental "manifestations" occur more frequently it will become more and more difficult to put the economic interests of the few against the survival of the many.

Despite the differences described above, the fact remains that global ecopolitics have changed substantially since the mid-1980's, especially after UNCED. The rhetoric now is that of sustainable development. Countries that depend on funding from institutions such as the World Bank have to adjust to new environmental requirements.   The Bank and other international organizations are reluctant to fund environmentally controversial development projects.

Even so, the dominant paradigm is still that of consumerism or predatory capitalism. The change in the rhetoric of global ecopolitics does suggest that we are in the early stages of a paradigm shift. Unless major changes in technology occur in the years to come, predatory capitalism may render itself extinct by ecological necessity. The increasing occurrence of ecological disasters will force us to reevaluate the system. By this we do not imply the coming of a new communist age. For us this is an unlikely probability. It is difficult to predict how human beings will react to climatic change. The world of the 21st century will not be able to disregard the environment as the world of the 20th century did. It is difficult to imagine a world in the next century in which the profit concerns of those who benefit from environmental destruction will prevail. It is more likely that those who benefit economically from capitalism will adjust to the new environmental realities because profitable industrialization requires natural resources. If capitalism survives as a system, it will be a very different system. It will be far from laissez-faire. Environmental needs will dictate that we shy away from its modern consummerist culture. According to economist Herman Daly, these changes will require more than the inclusion of environmental costs, which until now have been treated as externalities:

> The internalization of externalities is a good strategy for fine-tuning the allocation of resources by making relative prices better measures of relative marginal social costs. But it does not enable the market to set its own

absolute physical boundaries within the larger ecosystem. To give an analogy: proper allocation arranges the weight in a boat optimally, so as to maximize the load that can be carried. But there is still an absolute limit to how much weight a boat can carry, even optimally arranged. The price system can spread the weight evenly, but unless it is supplemented by an external absolute limit, it will just keep on spreading the increasing weight evenly until the evenly loaded boat sinks. No doubt the boat would sink evenly, ceteris paribus, but that is less comforting to the average citizen than to the neoclassical economist (Daly, 1977:69).

The economy proposed by Daly is one controlled by the state, one that puts social and environment needs above commitment to a free market. He questions the validity of neo-liberal claims that the market can set its own boundaries. Is a green socialist state the wave of the future? If one assumes current economic trends this does not appear to be the case. Countries throughout the world are participating more intensely in global capitalism by adopting liberal economic policies such as less government regulations. However, the deteriorating state of the environment may force us in the direction of further regulation of the economy. It may take only a few very severe environmental "manifestations" to force us in that direction. If these do not occur we will continue to take conservative steps towards a paradigm shift.

No matter how small the steps taken towards sustainable development in the 1980's and 1990's they had profound consequences for Brazilian ecopolitics. Global concerns about the state of the environment made Brazil an environmental villain for destroying the Amazon rainforest and for treating its native peoples so poorly. Accusations primarily came from First-World environmentalists, environmental organizations, and governments. The accusations reflected legitimate concerns about the way Brazil was handling the environment. However, from the Brazilian perspective these accusations were an attempt to meddle in Brazilian internal affairs, an attempt to violate its national sovereignty over Amazônia. It viewed the whole issue as a form of environmental imperialism, especially by the United States.

Notes

[1] The study also presents results for East Asia (including India and Turkey), Eastern Europe, Scandinavia, and other European countries, and Nigeria.

[2.] Corporations, for example, put a commercial in the network CNN accusing the treaty to reduce CO_2 emissions to be unfair to the developed countries. The accusation was based on the proposal that developed countries should cut their emissions at a higher rate than developing countries, to allow the latter room for development.

[3] It should be noted that this tradition of parallel forums dates back to the U.N. Conference on the Human Environment in Stockholm in 1972 (Union of International Organizations, 1995).

[4.] The speeches from world leaders were televised live in Brazil by the Brazilian Educational TV (TV E) during UNCED.

[5.] Translated back to English from the Portuguese. Author's translation.

Chapter 5

From Environmental Imperialism to Neo-Liberalism

As we saw in Chapter 1, Brazil almost overnight became an environmental villain when the ecopolitics of the world-system changed in the mid-1980's. The pressure from environmentalists and world leaders was not well received by the political establishment. The calls for preservation were viewed with great suspicion, at best as an attempt to interfere with Brazilian national affairs, and at worst as an attempt to internationalize Amazônia. Ignorant of Brazil's nationalism over Amazônia, foreign critics ignited very strong feelings among certain elements of the population, especially the military and the landed elites of the Amazonian states. For some Brazilian right-wing nationalists, environmentalists became enemies of Brazil. Foreign environmentalists were viewed as using the environment and the Indians as an excuse to internationalize the region, thus giving the rich countries control over its vast natural resources. Brazilian environmentalists and grassroots leaders were viewed as collaborators in this process. Some were charged with crimes against Brazil when they went abroad to speak on behalf of their cause. For these nationalists, Amazonian issues were to be dealt with by Brazilians, not foreigners. A statement in the newspaper *O Estado de São Paulo* is illustrative:

> This country has an owner. The good and the bad, the rights and the wrongs, the wealth and the Brazilian problems are ours. To us and to no one else belongs the administration of our business and destiny (March 1, 1989).[1]

The reaction of President José Sarney (1985-1989) to such international criticism was belligerence. Sarney belonged to the old guard of politicians attached to the military regime. His government was transitional, highly manipulated by the military who nevertheless wanted to remain in the shadows (see, Zirker, 1991; and Zirker and Henberg, 1994:264). Before becoming president, Senator José Sarney had been the leader of the military party, the *Partido Democrático Social* (PDS), in the Brazilian Congress. He was a civilian president surrounded by military officers who occupied important positions as advisors in his administration. During his term in office the threat of a coup was always present. The military used this threat to suppress civil society every time decisions seemed intolerable (Zirker, 1991:50-61).

The influence of the military over Sarney concerning Amazônia is illustrated by his decision not to attend an international conference on the environment in The Hague. Richard House of *The Washington Post* wrote on March 4, 1989:

> Sarney had accepted French president Francois Mitterand's invitation to attend the conference, organized by the Dutch. The meeting [which began March 10, was] expected to debate the creation of an international environmental group to monitor tropical affairs. . . . After a meeting with the chiefs of the Army and the military intelligence service, however, Sarney changed his mind . . . . Instead he issued a fresh attack on "intolerable" meddling in Brazil's domestic affairs. "We are masters of our destiny and will not permit any interference in our territory," Sarney said.

The *Post* also noted that 'Army minister Gene. Leonidas Pires Gonçalves . . . attacked "false ecologists," whose objective he said was to "internationalize Amazônia'" (see also Zirker, 1991:61)

For Sarney the criticisms against Brazil were "unjust, defamatory, cruel, and indecent." He argued that because of a staggering foreign debt it was unrealistic to expect Brazil to restrict its economic development over environmental issues. He was especially critical of the United States. He believed that industrialized countries such as the United States had no right to lecture Brazil on environmental responsibility when they were the biggest polluters (*Time*, September 18, 1989). When Sarney was asked about the Amazon by a group of visiting U.S. Senators, which included then Senator Al Gore, he had the following reverse accusatory response: "What about the Tongass?" He was referring to the U.S. policy of subsidizing the destruction of huge trees in the Tongass National Forest in southwestern Alaska (*San Francisco Chronicle*, October 1, 1989).[2] The message was clear: the U.S. should look at its own domestic situation before criticizing Brazil.

It seemed to the Sarney administration that the developed countries were using the environment to curtail the development of third-world countries, a form of environmental imperialism. The call for preservation was ironic for many Brazilians because it was coming from governments with a long history of environmental destruction and, in the case of the United States, a long history of violation of Indian rights. Brazilian officials claimed that the rich countries had used their natural resources to achieve very high levels of economic development. Now, it was the turn of third-world countries. Parts of a speech by former U.N. Ambassador Paulo Nogueira-Batista (1989) is revealing:[3]

> The heart of the matter lies most probably not in what has been proposed in the Bruntland Report as "environmentally sustainable development." We hope this may [not] be just an euphemism for subordinating progress in the Third World to a cleaner global environment, unilaterally defined by the industrialized countries. As a matter of fact, the global environment is certainly much more threatened by what has already happened and is happening in the industrialized world. Legitimate environmental concerns cannot, should not, be translated simplistically in formulations which may result in freezing at abnormal, subhuman, low levels the living standards of mankind's great majority (p. 20).

For Nogueira-Batista the development of third-world economies was a matter of human rights, recognized by the United Nations as part of the decolonization process:

> The right of all nations to develop came to be recognized by the international community as a natural consequence of the political decolonization process which took place after II World War. The independent under-developed countries were given to understand that to exercise that fundamental right, they were entitled to the same technological solutions used firstly by the market-oriented economies of the west and subsequently by the centrally-planned economies of the east (p.1).

Nogueira-Batista's speech conveys the fact that the industrialized countries provided the third-world with models of development. Suddenly, when there was a threat to their own survival in the possibility of drastic environmental problems such as the ozone hole and the greenhouse effect, the developed countries were attempting to obstruct the use of these models by the developing countries. To him, this was a violation of human rights. Why should a minority of the global population have access to so much when the majority went empty-handed and hungry? Cries for preservation, and especially impositions by organizations such as the World Bank were seen as a form of imperialism (Barbosa, 1990).

The statistical evidence spoke for those who claim environmental imperialism. The share of pollution spewed by the developed countries has been disproportionate to their population. For example, while Brazil's carbon emissions were 50.2 million metric tons in 1987, and for the whole of Latin America 229.7 metric tons, the United States alone emitted 1,224 billion metric tons. Of the estimated 5.3 billion metric tons of industrial emissions, North America, Western Europe, and Japan were responsible for roughly 45% of the total (2.4 billion metric tons). These figures do not take into account the cumulative effect of western industrial pollution. The developed countries have been polluting the atmosphere massively longer due to their earlier industrial revolutions. In 1987, 89% percent of the world's population contributed 20% of carbon dioxide emissions, while 11% of the world's population contributed over 65% (*Time*, January 2, 1989; Ruckelshaus, 1989:170). Alan Durning (1992) illustrates the income divisions that have dominated the world system:

> The world has three broad ecological classes: the consumers, the middle income, and the poor. . . . The world's poor--some 1.1 billion people-- includes all households that earn less than $700 a year per family member. They are mostly rural Africans, Indians, and other South Asians. . . . The poorest fifth of the world's people earn less than 2 percent of world income. . . . The 3.3 billion people in the world's middle-income class earn between $700 and $7,500 per family member and live mostly in Latin America, the Middle East, China, and East Asia. This class also includes the low income families of the former Soviet bloc and of western industrial nations. . . . The consumer class--1.1 billion members of the global consumer society-- includes all households whose income per family is above $7,500. . . . The consumer class takes 64 percent of world income--32 times as much as the poor . . . (pp. 26-27).

For the Sarney administration, and for Third-World leaders in general, the issue was clear: the call for preserving tropical rainforests was an attempt by the developed countries to protect their lifestyle. It was assumed that the developed countries wanted tropical ecosystems to serve as an absorbing mechanism, a sink, for their pollution:

It would be ludicrous to expect Brazil. . . to assume the burden of refraining from rationally exploiting its forests under any circumstances, at the same time taking upon itself the responsibility of actually increasing our forest coverage. All of that in order to reduce our own limited emissions of CO_2 and at the same time to help absorb the unchecked voluminous emissions resulting from industrial activities in the developed countries (Nogueira-Batista, 1989, p.16).

The nationalism was so intense that the Sarney administration responded to proposals for "Debt-for-Nature" programs by environmentalist organizations with great suspicion. When interviewed about this issue the President stated, "There is no amount of international money that can buy a square meter of Brazilian Amazon soil." For him such plans meant the "internationalization" of the Amazon. He added, "One thing that we cannot accept is to exchange our sovereignty or a piece of our territory for any type of economic aid or the foreign debt. This would be abdicating our sovereignty" (*San Francisco Chronicle*, February 11, 1989).

Sarney remained belligerent toward international public opinion to the end of his administration, but he took steps to appease it, moved by fears of losing international funding. On October 12, 1988, he created the *Programa do Complexo Ecosistema da Amazônia Legal* (Program for the Defense of the Ecosystem Complex of Legal Amazônia). This program became popularly known as *Programa Nossa Natureza* or Our Nature (*Informativo FFCN*, 1988; *San Francisco Chronicle*, October 13, 1988). Even though Sarney claimed that "the red light that awakened him, was the disclosure of Brazilian scientists that they had found more than 6,000 man-made fires in the Amazon in a single day" (*San Francisco Chronicle*, October 13, 1988), the truth about the program is that it was a nationalistic call to appease international public opinion, keeping the Amazonian issue for discussion among Brazilians. Its nickname, *Nossa Natureza* (Our Nature), is revealing of this.

Nossa Natureza was important in reducing Brazil's rate of deforestation, mainly by eliminating the tax incentives for development in the region. It also enacted stiff penalties for illegal burnings, and suspended some road building programs. In addition, it stipulated mercury control in mining and called for the mapping of "agro-ecological" zones, zones where soils could sustain agriculture. It created IBAMA out of four disconnected environmental agencies. The environment now came under the control of one agency devoted exclusively to it. The government also allocated U.S.$178 million dollars for its first year. By September 1989 the government claimed that it had levied U.S.$10 million dollars in forest burning fines (*San Francisco Chronicle*, April 7, 1989; and September 25, 1989).

According to data from the Instituto Nacional de Pesquisas Espaciais (INPE) shown in Table 2, the annual deforestation rate for the years 1987-1988/89 for legal Amazônia was 0.48%. The following fiscal year, 1989/1990, it declined to 0.37%; and to 0.30% in 1990/1991 (INPE, 1988, 1999; see also Worcman, 1990:48).[4] However, in 1992 rates increased again, reaching 0.81% in 1994/96. Better monitoring combined with a moratorium in the export of Mahogany reduced the rates to 0.51% for 1995/1996 and to 0.37% in 1996/1997.[5] Despite the oscillations the program did help combat deforestation. However, by August 1990, 415,200 km² of Amazônia had been destroyed. Thus, *Nossa Natureza* helped reduce the problem, not eliminate it.

It is interesting that Sarney walked a thin line between his traditionalism, the old-guard mentality about Amazônia, and his new-found ecologism to appease ecologists. For example, he wanted to create his own *obra faraônica*, his "pyramid," in line with previous Brazilian leaders. He revived a project to construct a 1,609 kilometer long railroad connecting Acailândia, along the Carajás-São Luís railway, to Anápolis near Brasília. But in line with the parochialism of so many politicians, the new railroad would substantially benefit the wealthy landowners of his home state of Maranhão. The project was also criticized for benefitting a few construction companies to which the government owned $5 billion dollars (Hall, 1989:60-61; Simons, 1987:G7). It is doubtful that without international pressure Sarney would have taken any steps towards environmental preservation.

International pressure alone was not the only reason for Sarney's environmentalist stance, however. Despite its strong ties to the old guard, his administration was a civilian government operating within the context of a transition to democracy. The dissatisfaction with what was taking place in the forest was coming from within Brazil as well. Indians, rubber-tappers, landless peasants and ecologists were not about to miss this moment in Brazilian history. They would lobby as much as possible the *Constituintes*, the sessions on the writing of the New Constitution. A free media also covered events within the country, exposing the plight of several groups to the Brazilian public in general. While, we shall deal with the situation of Indians and rubber tappers in Chapter 6, it is important to take note of the overall changes then taking place in the ecology movement in Brazil. It became a force to reckon with.

Table 2. *Amazonian Annual Deforestation Rates (%)*

State	77/79	88/89	89/90	90/91	91/92	92/94	94/95	95/96	96/97
Acre	0.42	0.39	0.39	0.28	0.29	0.35	0.86	0.31	0.26
Amapá	0.06	0.12	0.23	0.37	0.03	-	0.01	-	0.02
Amazonas	0.10	0.08	0.04	0.07	0.06	0.03	0.14	0.07	0.04
Maranhão	1.79	1.30	1.03	0.63	1.07	0.35	3.21	2.01	0.40
Mato Grosso	1.01	1.31	0.90	0.64	1.05	1.40	2.43	1.56	1.25
Pará	0.62	0.55	0.47	0.37	0.37	0.42	0.78	0.62	0.41
Rondônia	1.11	0.78	0.91	0.62	1.27	1.45	2.75	1.45	1.18
Roraima	0.18	0.39	0.10	0.27	0.18	0.15	0.14	0.14	0.11
Tocantis	2.97	2.00	1.61	1.61	1.17	0.95	2.29	0.94	0.81
Total	0.54	0.48	0.37	0.30	0.37	0.40	0.81	0.51	0.37

Source: Instituto Nacional de Pesquisas Espaciais (INPE) (1999)<http://www.inpe.br/>

A small environmentalist movement existed in Brazil even during the military years. The Brazilian Federation for Nature Conservancy (FBCN: *Federação Brasileira de Coservação da Natureza*) was formed in 1958. Also, leading Brazilian environmentalist José Lutzenberger founded the Gaucho Association for the Protection of the Natural Environment (AGAPAN: *Associacão Gaúcha de Protecão ao Ambiente Natural*) in June 1971. As a result of some political liberalization in 1974, a few ecological associations were formed in major Brazilian cities, e.g., the Ecological Art and Thought Movement (*Movimento Arte e Pensamento Ecológico*) in São Paulo. However, the movement was apolitical (Viola, 1988:214-218; Worcman, 1990:46).[6] Participants focused on denouncing pollution in major cities and the creation of alternative rural communities. This was anamolous in a country where environmental degradation had to a large extent occurred due to political decisions, but it was necessary in order to avoid persecution. When attempts were made to politicize environmental groups and/or demonstrations the apolitical elements of the movement intervened to restore its apolitical character (see examples given by Goldstein, 1992:141-142). However, as Brazil entered the period of increased democratic freedoms with *Abertura Democrática*, environmentalists became more outspoken about the need to participate in the political process, of being political. A statement by José Lutzenberger illustrates the trend: "The citizen is realizing that he needs to participate in politics because if not the bureaucrats will steamroll right over him. He needs to participate to know what is happening and he needs to shout, even if it is in vain" (*O Estado de São Paulo*, March 5, 1978, in Goldstein, 1992:143).

The 1980's marked the "ecopolitical" age of Brazilian environmentalism (Viola, 1988:214-218). Environmentalist organizations sprouted throughout the country and they were less reluctant to criticize the government.[7] By 1984 a survey placed the number of institutions at 900, with a total of 35,000 members (Goldstein, 1992:147).[8] One of the most active of these organizations was S.O.S. Mata Atlantica created in 1986. This organization has devoted itself to the protection of the remaining 5% of Atlantic forests of Brazil, a forest that occupied most of the littoral of the country when the Portuguese arrived in 1500. The importance of this organization is that it focused on preserving a forest located close to where most Brazilians live, in the urban centers of Rio de Janeiro and São Paulo, for example. The organization launched a strong campaign along nationalistic lines to preserve the forest, thus playing the same nationalistic advertising games that the military had used to promote development. TV commercials and T-shirts portrayed the green background of the Brazilian flag being eaten away with the slogan "They are tearing the green from our land." These efforts were successful in attracting financial resources from international organizations such as World Wildlife International (Goldstein, 1992:152-53; Worcman, 1990:51).[9]

The reaction of conservatives was to accuse NGO's of being controlled by foreign interests every time the status quo was challenged. The issue of funding was brought up by the conservative *O Estado de S. Paulo* newspaper in 1987, which claimed that Indian organizations were funded by foreign money channeled to Brazil via the Catholic Church (see Chapter 6). The fact that some of these organizations received money from abroad[10] made them suspicious for a government paranoid about the internationalization of Amazônia. NGO's defending the rights of Brazilian Indians during the *Constituintes* became special targets. The demarcation of Indian lands became a controversial issue during the process. The conservatives thought that Brazil was losing control of its territory to the Indians with the help of organizations aided by foreign capital (see Maybury-Lewis, 1989:3). In this dialectical process these organizations were often intimidated by being labeled "subversive," reminiscent of the dictatorship. However, intimidations did not work this time. Members of the old guard did not understand that democracy operates under different principles. The movement to save the forest was not going to disappear because NGO's were being labeled "subversive." NGO's could count on a free media through which to defend themselves from false accusations. They also counted on international allies to put pressure on the Brazilian government. Instead of retreating after authoritarian threats, the ecology movement in Brazil entered the ecopolitical age. It infiltrated the halls of government itself.

Of key importance to an understanding of the pressures put on the Sarney administration is the institutionalization of the green movement. Democracy allowed for a multiplicity of political parties, among them the *Partido Verde* (PV) or Green Party. On June 1985, a Provisionary Organizing Commission of the Green Party met in Rio de Janeiro. On October of the same year the party had its first convention, with delegates from nine states. The PV was officially founded in January of 1986. Due to stringent electoral laws, it was not recognized as an official party by the government. In order to participate in the elections the PV had to form coalitions with other parties. The *Partido Trabalhista* (PT) or Workers' Party in Rio de Janeiro was especially willing to enter into such an alliance, since it had not done well in previous elections. The PT-PV coalition for the 1986 election campaign was highly image-oriented and designed to capture as much free media exposure as possible (Goldstein, 1992: 170; Keck, 1992: 162; Viola, 1988:222-225). Media attention was easily drawn because the best known members of the PV were political dissidents of the Brazilian military who were allowed to return home in the political amnesty of 1979. This red-green alliance brought forward a new generation of green politicians who would keep environmental and social issues front stage.

In the 1986 elections in the state of Rio de Janeiro, the coalition chose green leaders Fernado Gabeira and Carlos Minc as candidates for governor and state deputy, respectively. Both candidates were ex-exiles, ex-guerrillas and thus the focus of media attention. Gabeira also embraced some very controversial issues, e.g., the legalization of marijuana, and he sought the support of groups overlooked by other political parties, e.g., gays. He also astonished conservatives by declaring himself "pan-sexual" and appearing at Ipanema beach wearing a minuscule G-string. The party also organized some of the biggest demonstrations in recent Brazilian history. The feminist demonstration "Speak up Woman," drew around 80,000 people, and the "Hug the Lake" campaign around 100,000 (Partido Verde, n.d). The idea of having people embrace a polluted lake in order to save it was novel in Brazil's political history.

Notwithstanding the attention drawn by Gabeira he received only 7.8% of the vote and lost the election (Partido Verde, n.d.). However, Carlos Minc was elected as state deputy, a significant victory for the environment. A principal focus of his campaign was transportation, arguing that motor vehicles were the greatest source of air and noise pollution in Rio. He and a group of Greens "descended onto a busy thoroughfare in downtown Rio, stopped traffic, and placed potatoes in the exhaust pipes of the black-smoke-belching diesel buses. To calm the annoyed motorists, flowers were distributed. . ." (Goldstein, 1992:173).

The number of political candidates elected under the PV is small when compared to other parties. However, the significance of these politicians for ecopolical changes within Brazil is enormous. From a dialectical perspective they introduced to Brazilian society a "post-materialistic," ecological view in opposition to the traditional "development-at-all-costs" view of the old guard. They tied the green movement to social justice and equality. The *Manifesto* of the Party (*Manifesto do Partido Verde*) begins with the following statement:

The Green Party (PV) has been formed to fight for liberty, peace and ecology, for human rights, for autonomy, self-determination and life alternatives. . . . The Green Party defines itself as a movement of citizens and not of professional politicians. . . . It is also part of a social political bloc that fights a broader battle against oppression, inequality, hunger, misery, and the predisposition of the elites towards corruption, cultural backwardness and other remainders of authoritarianism (Partido Verde, 1986).[11]

Of perhaps greater importance for Amazônia per se was the election of "ecological candidate" Fabio Feldmann as federal deputy officer, thus carrying environmentalism to the halls of Brasília itself.

Feldmann was involved with environmental issues since the 1970's. In the 1980's he founded his own environmental organization, OIKOS-Defenders of the Earth. In 1986 he entered the opposition party PMDM and campaigned as a defender of the environment. His campaign promoted the inclusion of a chapter on the environment in the New Constitution. He was elected with 46,000 votes and became the only federal deputy with an ecological platform out of a total of 559 deputies. His constituency was predominantly young, educated, wealthy people, particularly from the Jewish community of São Paulo (Goldstein, 1992:164).

As the only "ecological" deputy he was the butt of jokes and hostility from 'his fellow deputies and members of the Brazilian government. Graffiti on a wall by his house told of the reception of his ecological stance: "Jewish ecologist--get your hands off the Amazon." One military official also told him, "If you were in nazi Germany right now, you would be in the Fifth Column, and you would be dead" (Goldstein, 1992:164-165). He turned tables when he took fellow deputies to sites of environmental degradation, on "ecotours." His colleagues soon realized they could benefit politically from these events due to the media attention they received. The number of tours soared. As a result more deputies became affiliated with the Green Front in the Assembly, totaling some 90 members. Feldman's efforts led to the inclusion of Article 225 in the New Constitution of 1988, "...the most advanced constitutional treatment of the environment in the world, which asserts a healthy environment as a fundamental right of every citizen" (Goldstein, 1992:165).

The creation of the Green Party, the election of "ecological" politicians, and the inclusion of the environmental chapters in the New Constitution meant that it was no longer business as usual in Brazil. The cries for environmental protection coming from abroad echoed within the Brazilian green lobby, and vice-versa, forcing the government, despite its nationalist development stance, to pay attention. But, as we shall see in Chapter 6, the biggest uproar did not come from the urban centers; it came from the Amazon itself, from the "people of the forest." This time they were not going to be silenced. The world-systemic dialectical nature of the struggle over Amazônia becomes clear with their cause, and the reaction of the Brazilian government.

Collorism and Greenness

In the late 1980's the devastation in Amazônia had become a critical issue in Brazil's dealings with international organizations, especially the World Bank. International public opinion was also very clearly against Brazil's nationalistic, developmentalist stance over Amazônia. The country was also "against the wall" because it was in the process of renegotiating its mounting foreign debt, and attempting to obtain further loans for development projects. Thus, international public opinion mattered. It could influence the negotiations with the IMF and the World Bank. Despite the pressure, however, the Sarney administration chose to protect Brazil's sovereignty. It was "an-old-guard position." Brazil, they said, could not be criticized because the issue was a domestic one to be dealt with by Brazilians. Human rights issues such as Indian rights were simply a ploy by foreign governments through the maneuvering of environmentalist organizations to take over the region. Environmentalists, Indians and anyone involved in the environmentalist cause were to be viewed with suspicion, as enemies of Brazil.

The election of 1989 changed the rhetoric over Amazônia substantially. It brought to the center stage a young politician from the poor northeastern state of Alagoas, Fernando Collor de Mello, "Collor." Even though Collor was to be impeached for corruption in 1992, during the election campaign he captivated many Brazilians. Collor was different. He was young, energetic, handsome, educated, fluent in foreign languages, and an excellent communicator. Above all, he was the quintessential "a-la-first-world" media candidate, using every opportunity to project a dynamic image, especially through television. For example, he was frequently photographed while jogging, a different image in politics in a country where military presidents were old, and where the bulk of the population is young.

Collor's main opponent was Luís Ignácio da Silva, "Lula," a veteran union leader who led industrial workers in the state of São Paulo in strikes beginning in the late 1970's. With the legalization of political parties brought about by *Abertura Democrática* Lula became the strongman of the Worker's Party (PT). As a the PT candidate his agenda was one of social justice and equality, issues in urgent need of attention in a country where the majority of the population lives in dire conditions. Thus, Collor's victory is paradoxical when one takes into account that his political platform was far from being populist. Indeed, Collor was the favorite among the Brazilian elites. He posed less of a challenge to the status quo.

Since the 1950's Brazil had adopted the import-substitution model of development. This entailed government fomenting the growth of domestic industry and protecting it from foreign competition, usually with high tariffs on imported goods. This created a domestic industry accustomed to lack of competition, often producing goods of lower quality. Collor's agenda was to break this monopoly. He was a strong believer in privatization and market competition. He was also in favor of small government, a la Ronald Reagan and Margaret Thatcher.

One of Collor's appeals to Brazilians was his views (slogans) that Brazil was soon to become a developed country (see *Jornal do Brasil*, July 15, 1990), a vision of the future held by most Brazilians who associate the size and natural resources of their country to its potential for development. In this vision Brazil was to behave in syncronicity with the developed world to which it aspired to belong. Contrary to President Sarney, who accused the developed countries of environmental imperialism, Collor tried not to antagonize the First World. By addressing international concerns over Amazônia he found a way to maneuver the situation to Brazil's benefit. The rhetoric became an environmentalist rhetoric, that of sustainable development. When Collor visited Project Carajás, instead of emphasizing nationalism in the old way, he stated that "Brazil recognizes its mistakes in the environmental area." Even though he also affirmed that the developed countries have degraded their environments, he added that he was leaving ". . . Carajás confident that Amazônia [would] achieve its development while respecting its ecosystem" (*Jornal do Brasil,* July 15, 1990).[12] From the time Collor assumed office the ecological issue was a major concern of his administration.

One of the first steps taken by Collor to appease international public opinion was the appointment of Brazil's leading environmentalist, José Lutzenberger, as Special Secretary for the Environment. Even though Lutzenberger eventually antagonized environmentalists by his eccen-tricity,[13] this appointment was seen as a step in the right direction. It marked a transition in Brazilian environmental politics from military to civilian leadership, a feat that did not amuse military leaders. For them,

Lutzenberger did not represent the interests of the nation on such a critical issue as Amazônia.

Another symbolic step taken by the Collor administration was the use of Brazilian airforce helicopters to patrol Amazônia. These helicopters were to help the poorly funded IBAMA to control forest fires set by farmers. The step was largely symbolic because of the extent of the forest. IBAMA continued to be an underfunded agency (*Jornal do Brasil*, July 14, 1990).[14] However, it represented an attempt to integrate the military, to change its international image from villains to collaborators in the process of preservation.

The Collor administration also stressed the demarcation of Indian lands, the creation of extractive reserves, e.g., rubber tapping, and ecological parks. It protected more land in the course of two years than any other administration in Brazilian history (Inter-American Development Bank et al., 1992).

Perhaps the most important step taken by Collor to change international public opinion was his lobbying the U.N. to make Brazil the host for UNCED. In this effort he was successful. However, by bringing the world to Brazil to discuss global environmental issues, he made Brazil's environment the focus of increased international media attention. The heat was on, both within and outside the country. But, the strategy paid off. If Brazil was under the spotlight during the conference, so were the leaders of developed countries. During UNCED several countries pledged money for the preservation of Brazil's ecosystem. Even before the conference (West) Germany had pledged U.S.$151 million, and Canada U.S.$77 million to save coastal forests. During UNCED pledges from different countries amounted to U.S. $4 billion. Japan alone pledged U.S.$1.1 billion (*Jornal do Brasil*, July 14 and 17, 1990).

Thus, the brief Collor presidency marked an important transition in the rhetoric of the Brazilian government, from militant nationalism to some degree of ecologism and action towards preservation. Collor was more sensitive than Sarney to the changes taking place in the ecopolitics of the world system. Instead of confronting international organizations with claims of sovereignty over Amazônia he created an image of cooperation. This cooperation meant that he was capable of obtaining money from the industrialized countries for preservation. Unfortunately, his administration ended with charges of corruption.

The administration of President Itamar Franco proved to be a vacuum in terms of policies relative to the Amazon rainforest. While Collor had a definite leaning towards preservation, Franco did not take any further steps in the politics of preservation. After him, President Henrique Cardoso's record was dubious. Cardoso's economic agenda is even more of a neo-liberal agenda than Collor's. He is pro-business, pro-privatization, and a strong believer in small government. He has attempted to reduce the size of the Brazilian government by privatizing state-owned corporations. He has also moved Brazil to participate further in the world-economy, e.g., by opening markets. He has been accused by NGO's of attempting to reverse the gains made by Indian organizations, all for the benefit of business. On January 8, 1996 he signed Decree#1775 which allowed private interests and state and local governments to challenge the demarcation of Indian reserves (Borges, 1996). As we shall see in Chapter 6, this caused an uproar. Soon after the passage of this decree Brazil was on the defensive once again on an international level. As someone trying to make Brazil a major player in the global economy, President Cardoso had much explaining to do.

Notes

[1] Author's translation.

[2] According to *The San Francisco Chronicle* (October 1, 1989), an investigation by *The New York Times* found that the Forest Service had been selling 500-year-old trees for about half of the price of a cheeseburger. "Subsidies made it possible for pulp mills to buy Tongass lumber at about $2 per 1,000 board feet. . . . Comparable timber goes for $200 to $1,600 on the open market .

[3] To be fair it is important to point out that the transcript has a disclaimer. It reads as follows: "The views and concepts in the text, although based on the position of the Government of Brazil, are the exclusive responsibility of the author."

[4] INPE lists the creation of IBAMA, the Amazonian Monitoring Operation, and the restrictions on tax incentives as responsible for the decline of rates of deforestation from 1988 to 1991 (see INPE 1998, Figure One).

[5] INPE (1999) estimates a 10% increase in the overall rates for 1998, to 0.47%.

[6] Eduardo Viola (1988:216) argues that "Normally the efficiency of ecological struggles in this environmental phase is extremely low in terms of actual gains, but it is significant if we take into account the ecologizing effects on the mentality of qualitatively important elements of the population."

[7] It should be kept in mind that the 1980's was the age of NGO's in Brazil in general. A survey of 1,000 organizations operating in the country showed that the majority, 55%, were founded in the 1980's, another 35% in the 1970's, and only 8.8% in the 1960's (*Folha de São Paulo*, July 17, 1988, p. A-11).

[8.] Official listings published by SEMA (Secretaria Especial do Meio Ambiente or the Special Secretary for the Environment) lists a more modest figure of 503 institution for 1987 (Landim, 1988:58).

[9.] It is important to keep in mind that the colors of the Brazilian flag do have patriotic meaning. The green represents Brazil's vast forests.

[10.] The newspaper *Folha de São Paulo* (July 17, 1988, 1o. Caderno, A-10) reported that according to the Brazilian Central Bank Brazilian NGO's received U.S.$8.2 million in 1985. The Bank acknowledged that probably some money entered the country without its knowledge. The newspaper claimed that the bulk of the money came from Germany and Holland.

[11.] Author's translation.

[12.] Author's translation.

[13.] Lutzenberger was criticized for not facilitating a broader debate on the state of the environment. He was accused of focusing on narrow issues such as animal protection to the neglect of other issues. He also appeared eccentric to many Brazilians. For example, he made an issue out of the use of fertilizers in the lawns of his Brasília home. However, he really antagonized members of the Brazilian environmentalist movement when he came out in favor of controlled hunting of endangered species. The argument was that the revenues collected from the managed hunting of these species would pay for the costs involved in their preservation (*Jornal do Brasil*, June 29, 1990).

[14.] It is important to note that Brazilian environmental agencies were highly understaffed in this period, with a ratio of one person for each 26,209 hectares of protected land or forest (*Jornal do Brasil*, July 8, 1990).

Chapter 6

The People of the Forest Against International Capitalism

The term "people of the forest" invokes visions of native Brazilians foraging Amazônia for their subsistence. It probably seldom invokes images of the "other people of the forest." While Native Brazilians were the original people to inhabit Amazônia, other people also came to depend on it for subsistence. In addition to some 220,000 Native Brazilians, there is a population of rubber-tappers, 68,000 families in 1989 (Fearnside, 1989:387), and *Caboclos*.[1] Rivalries have always existed among these peoples. Indians have fought themselves prior to and after the arrival of Europeans, with major conflicts among them sometimes manipulated by Europeans.

Indians saw the *Caboclos* and rubber-tappers as intruders to their land. In turn, *Caboclos* and rubber-tappers saw the Indians as a threat and were frequently hostile to them. These rivalries have characterized the history of contact between these groups for centuries. The threat of the disappearance of the forest they depend on, however, meant that they would have to put their differences aside and form alliances, e.g., among Indians, between Indian and rubber tappers. They had a common enemy in the capitalist forces that were threatening to squash them. The alliances have to some extent paid off. Many Indian lands and extractive reserves have been demarcated in the country since the mid-1980's. The success of these alliances should be viewed from the vantage of changing global ecopolitics. These changes have affected the dialectics over Amazônia, both abroad and in Brazil. Indians, for example, had resisted intrusion to no avail for centuries. In the 1980's their resistance began to pay off. What changed was the global scenario, the ecopolitics of the world-system. While, in the past, people took economic expansion for granted, the new resistance emerged during a paradigm transition, the first steps toward sustainable development.

Centuries-Old Resistance

The age of conquest and colonialism created by the expansion of the European world-economy had five major consequences for the native peoples of the Americas: 1) annihilation; 2) slavery; 3) assimilation as subalterns; 4) proletarianization as cheap labor; and 5) internal colonization into ethnic enclaves, i.e., Indian reservations. Where cheap labor was needed, assimilation, slavery and/or proletarianization went hand in hand. Annihilation has been the preferred mode where resistance was intense and land desirable. Internal colonization has often been the result of compromises in which the conquered were granted some land, usually, but not always, of little importance to the conquerors, as in the case of Indian reservations in North America. This resulted from treaties which reflect defeat and/or compromise. In recent years, however, the creation of reservations is more nearly a reflection of the successful resistance of native peoples who sought to secure some of their ancestral lands against the greed of outsiders.

Brazil is no exception to the outcomes listed above. First, from the very beginning, assimilation via Christianity was the goal of the Portuguese Crown (see Novaes da Mota, 1995). Second, due to a shortage of colonists, christianization was supposed to create "productive citizens for the Portuguese Empire;" either as peons or slaves. Third, where resistance to conquest occurred, annihilation through war was the result; those Indians who survived became slaves working for Jesuit priests and other Portuguese settlers. Finally, in the twentieth century, Brazil entered the age of internal colonialism with the creation of reservations, first due to the need to relocate Indian people from areas included in development projects; and later as a result of the demarcation of ancestral Indian lands. These outcomes are not mutually exclusive. The introduction of reservations did not mean that genocide stopped in the countryside, as we shall see below. Outcomes reflect both official and unofficial policy. While the official policy in post-dictatorship Brazil has been the demarcation of Indian reservations, as required by the New Constitution, unofficially the policy of assimilation continues.

From annihilation to internal colonization, the result has been the dwindling of the Indian population. This population has been reduced to roughly 0.2% of the original population, or 220,000 thousand. In the 20th century alone 90 of Brazil's tribes disappeared completely, while others, e.g., the Urueu-Wau-Waus[2] and the Kreenakarore, were substantially reduced by contact (Paiva and Junqueira, 1985:18; The New York Times, January 2, 1994; Ghazi, 1994). When the Kreenakarore were first contacted in 1972 they numbered 350, but:

> By January 1975, this number had fallen to 79, all of whom displayed clear signs of tuberculosis. . . . By 1974, they had become beggars, wandering along the new road, prostituting their wives and daughters, drinking alcohol, dirty, fighting over left-overs. In 1975, only three of them were over 39 years old. At this point, they accepted the invitation of the Txukarramei, their traditional enemies, to live with them in the Xingú National Park . . ." (de Souza Martins, 1990:257).

In his *Amazon Journal* (1997) Geoffrey O'Connor relates a similar exposure for the Yanomamis after contact with gold prospectors in the latter 1980's and through the 1990's:

> "In May I made fifty tests for malaria and thirty-one were positive," says Florence [a missionary nurse], wiping sweat from her brow as she shuffles about her tiny one-room clinic looking for a needle to give another injection to another Indian in search of more of the white man's medicine. "This was only in the month of May." She pushes down on the hypodermic, thrusting the needle into the arm of an old woman who flinches, cringes, and then lets out a squeaky little sign. "This is truly the year of malaria" (p.200).

In addition to malaria, the Yanomamis were afflicted by measles, syphilis, and other diseases. It is estimated that between 1987 and 1989 alone, some 15% of the Yanomami died from disease or malnutrition (UN Chronicle, June 1993, p.47; see Davis, 1993:36 also). By 1990 the devastating effects of malaria had captivated the international media and was a source of embarrassment for the Brazilian government (see O'Connor, 1990: 229).

Indianness and Brazilian Social-History

A curious feature of Brazilian social history, especially the ethnic relations literature, is the dearth of serious discussion on Native Brazilians as "an ethnic group" until the 1980's. It is true that Indians were described in literary works and even serious sociological studies. One of the most famous Brazilian sociologists, Gilberto Freyre, refers to the contribution of "Indian blood" to the Brazilian stock (see Freyre 1956). However, literary works tended to romanticize Indians as "noble savages"[3] and sociological works devoted scant space to them, often describing them as a disappearing stock of the Brazilian mixed heritage. The sociological literature, instead, focused on the African Brazilians. This bias reflects the importance that slavery had in the evolution of Brazilian social history. As the Indian population declined along the coast, where the bulk of the colonial population was, slaves replaced them as the preferred form of labor. With the exception of a few enclaves that managed to survive to the present along the coast, e.g., the Shoko and Kariri-Shoko and the Patoxó Hã-Hã-Hãe of Northeast Brazil (see Novaes da Mota, 1995; and da Silva, 1983:62), the Indian population of this area was obliterated. Those who managed to survive, such as the Pataxós, maintain a close relationship with the regional societies (Brazilian Embassy, 1993:4); that is, they are incorporated in the economy. While the African element was visible in most major coastal cities, the Indian element was not. It was "hidden" in the jungles of Amazônia. Therefore, it was easy for Brazilian society and scholars to ignore them. Indians became associated with far-away places for most Brazilians, places where, except for those in Manaus and Belém, few would ever visit.

Twenty one years of military dictatorship added to the obscurity of Indian ethnicity. The ideology of the state was that Brazil was "a racial democracy." Racism as an issue was swept under the carpet. This ideology was favored to explain the supposedly harmonious relationship between blacks and whites. But, as a bulletin of the Ministry of Justice indicates, Indians were not ignored:

> Nobody hates the Indian because he is an Indian. The antagonism emerges in the moment in which two cultural frontiers meet each other, when two ways of thinking collide and repel each other because of different values. This makes the white pioneer see in the Indian a stranger, an inferior being, like a poorly reared child with bad habits who can become aggressive and dangerous (Jobim, 1979:8).[4]

This quote blames antagonism on cultural differences rather than race: Not biology, but culture. In this view, if Indians were to have a culture that equaled that of other Brazilians the "antagonism" would disappear. In this way prejudice is cushioned. People can view themselves as not "really racist." Unfortunately, defining Indian culture as inferior to Brazilian culture, to be replaced by it, has given license to kill.

In Brazil, government's FUNAI became the determining agent of Indianness. Unlike the United States where being Indian has been determined by blood, in Brazil Indianness has been a question of acculturation. Unacculturated Indians were wards of the state, while those who lost their cultural identity were considered full citizens, and treated as non-Indians regardless of blood. In addition, the notion of "Indian" as an ethnic category has been alien in Brazilian society. Indians have been part of the Brazilian imaginary as "Brazilians" in another stage of development. The fact that the Indians themselves have referred to Brazilians as "the Brazilians," as foreigners, has been ignored by the authorities. Most Brazilians have also lumped all Indians together, under the category "Indian," a social construct based on ignorance of the diversity that exists among Brazil's Native peoples. There are or were 180 known indigenous groups in the country speaking 170 different languages (Brazilian Embassy, 1993:14). The Indians, of course, have referred to themselves as members of tribes. However, from the 1980's on, in an effort to unify themselves, they began to adopt the broader notion of "Indianness," of being *"Índio."* This notion, constructed by the outside world and internalized by the Indian, is illustrated by O'Connor (1997) who reports the Indian meeting in Altamira, Pará, in 1989:

> Around wooden tables they sit huddled in distinct ethnic clusters--representatives of the Xavante, Kayapó, and Panara people. There doesn't seem to be too much mixing among them, which is perhaps an indication of language differences and their own cultural barriers rather than any history of conflict between their societies. The intertribal battles of the past have dissipated in recent years as, one by one, these groups have come to recognize themselves as *Indios* and started to ally themselves in opposition to the national society of Brazil. Slowly they have begun referring to other Indians as their "relatives" while still characterizing most whites as a group not to be trusted (pp. 124-125).

The construction of this ethnicity is something new in Brazilian history. It corresponds closely to the image created by environmentalists and the media, an image that approaches the noble savage of the first contact with Europeans. In this 20th century vision of the noble savage the Indians are depicted as "protectors" of the forest, "saviors of earth," holders of knowledge of the forest. As we shall see, this image has become part of the "new resistance." The "old resistance" was less complex.

Old Methods of Resistance

Indians have resisted invasion as best they could over the centuries. John Hemming's *Amazonian Frontier* (1987) is an excellent account of this history of resistance. Throughout Brazilian history Indians have resisted enslavement and assimilation by attacking and killing the invaders. They, however, were also faced with the lure of the "white man's world." Resistance and acceptance of goods from Europeans and Brazilians helped determine the degree of assimilation of these people. In the old resistance Indians used the bow and arrow and/or moved as far as possible from non-Indians, deeper into the forest.

When Indians chose to confront non-Indians, their resistance prior to the 1980's was entirely parochial. It was limited to attacks using bows and arrows. These, however, should not be underestimated. In the fast-paced construction years of the 1960's and 1970's Indians were often successful in halting road construction temporarily by attacking construction crews. For example, in the late 1960's the Waimiri-Atroari resisted the construction of BR-174 through their territory by killing dozens of highway workers. The situation got so bad that on January 8, 1975 there were talks of using tear gas and machine guns to intimidate them because they had killed 15 government employees and road construction personnel in the previous year (O'Connor, 1997:177; Howe, 1975). Even the Urueu-Wau-Waus, who were almost obliterated by disease were fierce defenders of their territory (see von Puttkamer, 1988). In some areas Indian resistance so delayed projects that they were completed only when military force arrived (Bunker, 1985:118). The lack of freedom by the Brazilian press to document these events meant that in the end the Indians lost. The outsiders had guns, lure, and diseases. Indians had only their bows and arrows and their will to survive.

New Methods of Resistance

The military strategy of resistance has not disappeared among the Indians. It is common to hear reports that settlers have been attacked after invading their territory. And, people do get killed. These killings are often committed as revenge on the part of Indians, and vice versa. One of the most publicized cases in recent history occurred in August of 1993, when some 12 Yanomami were massacred by gold prospectors. The massacre occurred when the brother of a gold prospector who was killed by the Yanomami in a previous conflict, convinced ten of fifteen other gold prospectors to revenge his brother (*Veja*, August 25, 1993, p.24; see also O'Connor, 1997:356-359; and Weiss and Weiss, n.d., p.4).[5]

Sporadically, Indians have held settlers and public officials hostages as a strategy of resistance. The Indianist Missionary Council (CIMI) announced (Worldwide Web, November 13, 1992) that 82 persons (initially more than 100) were being held hostage by the Guajajara Indians of the state of Maranhão who demanded the removal of the village of São Pedro dos Cacetetes from their territory and the punishment of the persons involved in the murder of a 20-year old Guajajara Indian (CIMI, 1992a). On March 14, 1994 600 Indians (Makuxi, Ingarik, Wapixana, and Taurepang) set up two road blockades at the villages of Machado and Olho d'agua, in Roraima, demanding the demarcation of the Rapôsa-Serra do Sol Indian reservation. They demanded that gold miners be removed from their land and that their constitutional rights be respected by the Brazilian government (ProRegenwald, 1994).

What makes these cases so illustrative of the new strategies is that they were reported in the Worldwide Web. The Indigenous Council of Roraima (CIR) placed a report through a German (ally) provider, the ProRegenwald organization, asking for international support. It included the addresses and the telephone numbers of the Brazilian Minister of the Environment and Amazônia, and the Minister of Justice (see Pro Regenwald, 1994). A report was also placed on the Web by CIMI (CIMI, 1994a). This case shows how resistance has changed. The reports by CIR and CIMI, along with other accounts of events in Amazônia on the Worldwide Web (see Gaia Forest Conservation Archives), show that Indians understand Brazil's susceptibility to international pressure, and that international allies can play on this vulnerability to change policy.

The new strategy of the Brazilian Indians involves participation in the dialectical process over Amazônia. This strategy emerged with *Abertura Democrática*; that is, it flourished because of the civil liberties brought about by democracy, which allowed Indians to organize and form coalitions with environmentalists and grassroots organizations within Brazil and abroad. Central to this change is the participation of a free Brazilian media eager to cover events taking place in Amazônia, sparked to a large extent by national and especially international interest over the fate of the forest.

Abertura Democrática allowed distrust of the Federal Government to be ventilated in the open. FUNAI became a special target of this mistrust because of its constant failure to fulfill its promises. It was clear to those involved in the cause that the agency was a pawn of big business and that it did very little to defend Indian interest. This frustration, coupled with democracy, created an Indian phenomenon.

The Political Rise of Mário Juruna

Chief Mário Juruna of the Namunkura of Mato Grosso state is a unique character in Brazilian history: he was the first Indian to occupy a congressional office. The story began in 1977 when Juruna, fed up with broken promises, made headlines in Brazil and abroad by using a tape recorder to parade all the promises made to Indians by FUNAI representatives and other government officials. In a society in which all social groups had been gagged by the dictatorship, the idea of an Indian tape-recording government officials was an irresistible enticement for the news media. Juruna received a great deal of attention by reporters who were themselves beginning to relish freedom, and surprised by the idea of political participation by Indians. Juruna quickly became the best known Indian in the country (Hohlfeldt and Hoffman, 1982; Paiva and Junqueira, 1985:39). His use of the national media spearheaded a whole Indian movement on a national scale. He then used the media's attention to launch his own campaign for a seat in the Brazilian Congress.

In September 1981 he joined the *Partido Democratico Trabalhalista (PDT)* (Democratic Workers' Party) in the state of Rio de Janeiro. The choice of Rio, instead of his native Mato Grosso, is revealing. In Rio he found a more liberal, better educated constituency than in Mato Grosso, where elite interests governed forest devastation. His choice of the PDT had much to do with his friendship with anthropologist Darcy Ribeiro who was active in the party, a senator, and a defender of Indian rights. It had more to do with the party's interest in Juruna than with Juruna's commitment to the party:

> I am fruit of the earth and I want to defend my people's and the Brazilian people's interests. Nobody controls me, nobody influences my opinion. I am experimenting with the PDT. If this does not work out I will join another party (*Correio do Povo*, April 10, 1981, p. 15).[6]

For the Brazilian people Juruna's candidacy had an element of joke about it; after all he was an Indian playing politician in a political system of which people were tired due to corruption. However,

Little by little what for the majority of whites was a joke, a fad, became something serious. Even FUNAI, by attempting to manipulate Mario to its advantage in land disputes in the Alto Araguaia, got frustrated (Hohlfeldt and Hoffmanm, 1982:15).[7]

The truth is that Mario Juruna was serious about his cause and he was not for sale. His mission was to *"colocar a boca no trambone"* or to blow the horn about the Indian situation in Brazil:

Already in my first speech in the Chamber of Deputies I stated my mission: I came to defend my people's and the Brazilian people's rights at all costs, no matter whom it hurts (Juruna, 1986:34).[8]

Juruna was very open about his accusations. He did not even spare President João Batista Figueiredo when the general criticized his election to office. He reprimanded the president in a speech to the Chamber, a reaction allowed only because of the climate of tolerance created by the onset of democracy:

I would like to further protest the words spoken by General Figueiredo when referring to my election through the state of Rio de Janeiro. He criticized Rio's people for not knowing how to choose their political representatives for this Chamber. I protest. I do not see in the president authority to criticize 31,000 voters who voted for me for Federal Deputy, especially when he was elected by a single vote, that of his colleague General Geisel. The general president should utilize his precious time and political space to take measures in favor of the suffering Brazilian people, who are unemployed and hungry due to his disastrous and incompetent political and economic policies (Juruna, 1984:16-17).[9]

His attacks were specially strong against FUNAI which he thought to be corrupt and against Indian interests. It was an institution he truly disliked. He credited its persecution against him as his motivation to launch his campaign for Congress. He made the decision after attending the Bertrand Russel Tribunal in Holland in 1981. FUNAI and other government officials did everything possible to prevent his participation in the Tribunal, aware that he would have an international forum to expose Brazil on the issues of human rights and the environment. Juruna had to fight in court to obtain permission to leave the country. In the end he won. However, while in Holland, FUNAI attempted to use a strategy of "divide and conquer" by questioning Juruna's leadership among other Indian leaders. It questioned his right to lead and argued that he was growing too powerful. When Juruna returned to Brazil he was upset and ready for more political involvement:

> Frustrated by FUNAI's persecution against me I then went to Rio de Janeiro and there I became a candidate for the position of Federal Deputy. So many people laughed at me and they did not believe that I stood a chance. I did not give up and decided to fight, always remembering in my public appearances that the whites had an enormous debt with the Indians and that it was about time they respected the rights of Indigenous people (Juruna, 1986:33-34).[10]

In office, Juruna would do his best to change FUNAI's structure. The organization was then controlled by the military. The few Indians working for it were janitors and cooks. There were no Indians in positions of management within the organization (Juruna, 1986:35).

One of Juruna's main achievements as an Indian leader was the creation of the Union of Indigenous Nations (UNI) in 1980. Indians of different backgrounds had earlier come together to discuss their problems. In April 1978, for example, the First Assembly of Indian Leaders was held under CIMI's leadership. But UNI was "an Indian organization," not a catholic or other organization fighting for the Indian cause (Carneiro da Cunha, 1987:44-45; CNBB, 1986:16-17; Paiva and Junqueira, 1985:35; Ramos, 1997). It fused two Indian organizations. The first was a pan-tribal student association addressing problems faced by the Indian students living in the capital city of Brasília. The other was the embryonic Indigenous People's Federation of the state of Mato Grosso, a project sponsored by Indian representatives of several villages during the Indian Week celebrations in Campo Grande, Mato Grosso do Sul.

UNI's purpose was to: a)represent participant indigenous nations and villages; b) promote the return, legal possession, and further demarcation of Indian lands, and their exclusive use by Indians; c) provide Indians and their communities with knowledge of their legal rights; and d) promote cultural projects of community development. In 1981, the first congress held by the organization brought together 228 Indian leaders from 46 nations. The second, held in 1982, brought together 450 leaders from 50 nations (Paiva and Junqueira, 1985:38).

It is important to note that the creation of UNI was opposed by FUNAI which considered itself the sole legitimate representative of Indian issues. The administration of General João Batista Figueiredo (1979-1985) was also in opposition. The president's views were based on a report by the Brazilian Bureau of Investigation (SNI) which argued that UNI was dangerous and subversive. It claimed that Indians could easily be manipulated by people "who would be inclined to stimulate opposition against FUNAI" (Paiva and Junqueira, 1985:38; Pereira Gomes, 1988:210-212; Ramos, 1997).

In the early 1980's UNI joined hands with other Indian-rights organizations, such as the Comissão Pró-Índio (Pro-Indian Commission) of São Paulo against the passage of Decree No. 88.985, which would have opened Indian lands to mining operations. The Comissão collected 2,000 statements against the decree from around the world. It also launched a legal campaign against it, questioning its constitutionality. Despite this opposition, President João Figueiredo signed it into law on January 9, 1985. However, there was such an outcry that the very next day he revoked his decision (da Silva et al., 1985:8-9).

While in office Juruna also introduced legislation for the creation of a Parliamentary Indian Commission, whose purpose was to evaluate all legislation dealing with Indian issues. Despite the presence of several hostile politicians, by 1987 the commission (created in 1983) had been mostly favorable to the Indians (Carneiro da Cunha, 1987:31-45).

Despite Juruna's accomplishments, he was unsuccessful in his run for a second term in office in 1986. He received only 10,000 votes (Documentação Indigenista Ambiental, n.d.; Pereira Gomes, 1988:213). It is interesting that he was very upset when other Indians also ran for office and criticized him. It was as though that challenged his (traditional) authority. He accused some candidates of not being "real Indians:"

> Who are Alvaro Tukano, Marcos Terrena, Mtsu Bara Kadiweu, David Terena, Tapuia, Idjarrubi Karajá and others who are constantly attacking me in the press? They are Indians who never suffered with their people and lived their entire lives in cities. They never felt on their skins the suffering of their people, they do not know their traditions and for this reason should not speak for the Indigenous people they say they represent. They live by lecturing at universities (Juruna, 1986:33).[11]

Juruna contrasts the background of his opponents to his own grassroots background:

> My fight began in the village, against farmers, against the police, and against FUNAI and anyone who did not respect my people. Before I became a deputy I was a warrior chief, defender of my people. In my contacts with the invading society I discovered that the authorities were fooling us. This is the reason why I started to use a tape recorder so that unfulfilled promises could be exposed. Then, I became known nationally and internationally (Juruna, 1986:33)[12]

Juruna's failure to remain in public office did not lessen the intensity of the Indian movement. By the mid-1980's Indians were fully involved in the political process through action at the grassroots level (see Pereira Gomes, 1988:214). Among all Indian nations, the Kayapó were the most active.

Kayapó Resistance

In the mid-1980's the Kayapós emerged at the forefront of the Brazilian Indian Movement. The Kayapós number 2,208 people living in 14 villages in the Xingú Park and the areas of southern Pará and eastern Mato Grosso (Fisher, 1994:221; Citations from the Global Prayer Digest, 1984).[13] They are known for being a fierce people with a strong warrior tradition. In recent years they have been outspoken about several projects in Amazônia, especially the damming of extensive portions of the Xingú river. What is interesting about Kayapó resistance is that it has combined traditional warrior practices with manipulation of the media and national and international ecopolitics. It has also included attempts to gain control of the exploration of natural resources within their territory. They quickly learned that wealth gave them access to modern commodities, and power to determine the pace of incorporation. Led by the fierce Colonel Pombo (Colonel Pigeon) they embarked on a form of resistance that combined the old and the new.

In the mid-1980's Colonel Pombo personally commanded 35 warriors from Kikretum village against the gold mining operations in the Bateia area. They successfully cleared the area of 300 gold prospectors.[14] They took possession of the mining equipment and engaged in negotiations with FUNAI to obtain funding to keep production running. FUNAI was uncooperative. The Kayapós then followed a different course of action. They began to negotiate directly with gold prospectors and mining companies. Of these negotiations the most important was with the Stannum-Shelita mining company. On April 22, 1982, in the town of Tucuma, Colonel Pombo signed an agreement with the company, and registered it in a public notary's office. This was a first for a people that legally were still wards of the state:

> "...this document established the conditions for the exploration of two mining operations [*garimpos*] for the next three years: it limited the initial operation to the presence of 200 gold prospectors in the area, with the limit of 400 at any time; it required royalties of 5% of the mechanized production (deposited in a savings account in the CEF [Caixa Economica Federal] in an account

belonging to Col. Pombo; 10% of the manual production to be paid in cash (da Silva et al., 1985:123).[15]

The gold prospectors in the region had until the end of July of that year to link themselves to this system or leave the area. Kayapó worriers under the supervision of Col. Pombo would patrol the area.

Conflict between the Kaypós and the mining company soon emerged. The Indians grew increasingly suspicious of the miners whom they accused of cheating. Much of the payments were in kind. Airplanes full of meat, bottled sodas, bread, etc., were given to them as payment. Payment was also provided in the form of assistance for medical bills in the hospital in Tucuma. At one point in mid-1982 the situation grew tense, with the Indians increasing their patrols of the area and expelling several miners. Meanwhile Col. Pombo struck another deal with a miner for the exploration of a cassiterite mine. This agreement also called for payments to the Kayapós of 10% of production (da Silva et al., 1985:123-124).

In October 1982 Col. Pombo flew to Brasília to convince members of FUNAI of the benefits of these contracts. FUNAI was already suing to stop the transactions. The trip was to no avail. But FUNAI was unable to stop the Kayapós who were determined to remain in business. However, the lawsuit made operations too turbulent for Stannum-Shelita. The company had to interrupt production several times, and, in 1983, it withdrew. This infuriated Col. Pombo:

> Colonel Pombo did not like this. The regional FUNAI delegate escaped being killed because in the last minute he decided not to attend a technical delegation that visited the village Kikretum in the month of May (da Silva et al., 1985:125).[16]

The Kayapó continued to make deals with gold prospectors for a profit with the help of a former FUNAI local official who married a Kayapó woman.

In terms of civil disobedience, the most famous Kayapó demonstration took place in the town of Altamira, Pará in 1989. It was organized by Kayapó leader Paulinho Payakan as the First Meeting of Indigenous People of the Xingú. Its purpose was to protest the construction of a series of hydroelectric dams along the Xingú River. It brought together 600 Indians and numerous people from throughout the world, especially media people covering the event. In Altamira the Indians gathered in traditional Indian attire, chanting war cries and dancing in groups. When a Brazilian official attempted to explain the benefits of the dam he was confronted by a Kayapó woman who held a machete to his face while expressing her anger in

Kayapó. While there was an element of spontaneity to the demonstration much of it was orchestrated. Altamira was ". . .an attempt to win over the media, and hence the Brazilian and international public, in their efforts to block the dam construction project" (O'Connor, 1997:104). And, the Kayapós were masters in playing the media's game, providing it with images of "the natives" so sought at the apex of public interest in rainforests in the late 1980's and early 1990's. And they had a very famous ally.

The Rock star, Sting, took on the rainforest cause by associating himself with the Kayapó. He raised $1.2 million dollars for the demarcation of their reservation. He also created the Rainforest Foundation to gather money for the cause (see Geoffrey O'Connor, 1997:219; *Veja*, April 28, 1993).

This relationship with Sting lasted until 1993, when it became public that the Kayapós had made contracts with timber and mining interests to operate in their territory. This disclosure led to accusations on both sides, Sting accusing the Kayapós of using him:

> They [the Indians] try to fool you all the time and tend to see whites as a source of resources rather than a friend. . . . I was very naive and I thought that I could save the world selling T-shirts, but I achieved little (in *Veja*, April 28, 1993:74).[17]

In return, Chief Raoni said:

> I did not know that Sting was bad mouthing the Brazilian Indian. He earned a lot of money on my back. The Brazilian Indian does not need him. It is better that we forget Sting. Let him say whatever he wants (in *Veja*, April 28, 1993:74).[18]

Veja called this abrupt ending *"O fim do romanticismo"* or "The End of Romanticizing;" that is, the end of romanticizing the Indian as the noble savage, as the eternal protectors of the forest.

But, Sting was not the only international ally the Kayapós had. Prior to the meeting in Altamira, in 1988, a tour abroad of Kayapó leaders was organized by Friends of Earth and Survival International. Kayapó leaders Payakan and Kube-i travelled to Holland, Italy, Germany, Belgium, Canada, and the United States to meet representatives from government, multi-lateral development banks, commercial banks, tropical hardwood importing industries, and the public (in lectures) (see Fisher, 1994:222). The aim was to stop the projects they protested against in Altamira. These trips substantially benefitted the Kayapós because of media interest, thus putting pressure on public officials to bring about change in policy.

Another modern weapon used by the Kayapós has been the video camera. Their use of the camera goes beyond Juruna's use of the tape recorder described above. Their "appropriation of video" began in 1985 when they obtained their first camera. It has included not only the recording of political events, such as the events in Altamira, but also the recording of their own culture in films for themselves and the outside world. These were films produced by the Kayapó themselves and were largely "home movie" quality, even though they have been shown to audiences at film festivals (Turner, 1992:5-7). What is important about this appropriation is that the Kayapós were very aware of the media's interest in their use of the video camera; that is, "primitives" using modern technology. Anthropologist Terence Turner, who has worked with the Kayapós in their efforts at video production, says of this strategy:

> There is a complementary side to this point, which is that for a people like the Kayapó, the acting of shooting with a video camera can become an even more important mediator of their relations with the dominant Western culture than the video document itself. One of the most successful aspects of the series of dramatic Kayapó political demonstrations and encounters with the Brazilians (and other representatives of the Western world system such as the World Bank and Grenada Television) has been the Kayapós' ostentatious use of their own video cameras to record the same events being filmed by representatives of the national and international media, thus ensuring that their camerapersons would be one of the main attractions filmed by the other crews. The success of this ploy is attested by the number of pictures of Kayapó pointing video cameras that have appeared in the international press. The Kayapó, in short, quickly made the transition from seeing video as a means of recording events to seeing it as an event to be recorded (1992:7).

An excellent illustration is a photograph that appeared in *The New York Times* of February 25, 1990, p.E5, of a Kayapó Indian posing with a western reporter. They were photographed filming side-by-side, in the same pose.

The combination of strategies certainly produced results for the Kayapós. After the demonstration in Altamira the World Bank announced that it was denying the Brazilian government the loans to its power sector. Analysts hailed the event as instrumental in fueling a pan-Indian identity in Amazônia (Fisher, 1994:1994).

Even though the Kayapós became the main Indian activists against the implementation of development projects in Amazônia (Fisher, 1994:228), they were not alone. The Indian movement had indeed become pan-tribal. As noted, demonstrations during the *Constituintes* and in Altamira involved the presence of Indians from different tribes. The tactics of seeking international allies and international media attention were also not unique to them. Yanomami leader Davi Kopenawa also traveled to the U.S. in 1990 and made appearances in U.S. television shows, such as NBC's *Today* and the CBS evening news. He was profiled in the *New York Times* (O'Connor, 1997:232-256). There was a sense of urgency in his visit. Brazilian gold miners invading Yanomami lands brought with them a wave of malaria. The Yanomami were lobbying not only for the demarcation of their reservation but also for the expulsion of these people from their lands.

The Other People of the Forest

The rubber-tappers have been the other people of the forest most engaged in the grassroots movement to save Amazônia, and thus their livelihood. As already mentioned, in 1989 there were 68,000 families engaged in this type of activity in the region, a figure disputed by the rubber-tappers themselves who argued that many families went unaccounted (Fearnside, 1989:387). These people were descendants of those who began migrating to Amazônia from the Northeast in the 19th century. Like the Indians, they depend on large tracks of forest for their livelihood. And, like the Indians, they saw their way of life threatened by big landowners who came to the region with the advancement of the Brazilian frontier.[19]

Rubber tappers have been a disenfranchised category in Brazilian history. Throughout Amazônia they have worked in semi-slavery in the system of *aviamento*. The *aviamento* is a system of exchange in which a supplier, the *aviador*, provides owners of *seringais* (a *seringal* is a rubber-producing area owned or leased by a single individual), or *seringalistas*, with basic production and subsistence goods at the beginning of the season in exchange for payment in latex (Bunker, 1985:xiii, 66-67). The *seringalistas* in turn exchanges these goods with the rubber tappers. Thus, "The *seringalistas* [have] controlled both the tappers's exchange of rubber and their provision of subsistence goods and tools" (Bunker, 1985:67). The result of this system was that some tappers were so much in debt that they were in semi-slavery (Hecht, 1989:16). The abolition of this system of exploitation became a main goal of their grassroots movement.

The rubber-tappers' movement emerged in the state of Acre in the late 1970's. Its leader was Chico Mendes who literally gave his life to his cause, being assassinated on December 22, 1988. Mendes was the major force behind the creation of the National Council of Rubber Tappers' in the mid-1980's, which aimed at destroying the *aviamento* system as a whole:

> Mendes challenged more than debt peonage. Through collective action, Mendes and the tappers fought for the right to sell when and where they chose, to farm for their subsistence if they chose, and not to pay rent for the *seringalista* whose ownership rested more on coercion and market control than on legality. By asserting the rights of labor and human dignity over the claims of property, he drew battle lines that assured the enmity of landowners and ultimately his own violent death (Hecht, 1989:16).

Another threat lurked on the horizon: the opening of Acre itself as a new frontier.

In the early 1970's the government of the state of Acre led by governor Wanderley Dantas invited farmers in the rest of Brazil to "Produce in Acre, Invest in Acre, Export via the Pacific," an invitation mirroring the federal government's own plans for Amazônia. As a result of this initiative more than five million hectares changed hands between 1971 and 1975, equivalent to one third of the surface area of the state. As elsewhere in the region the land was being expropriated by cattle ranchers, even though in many instances cattle ranching was a facade for land speculation. Acre was also the destination of thousands of small farmers in search of a better life. This often led to bitter disputes when they expropriated land already owned by absentee owners. The violence resulted in massive out-migration. Approximately 10,000 rubber-tapper families crossed into Bolivia between 1970 and 1975, while the state's urban population increased from 59,307 to 81,970 (Bakx, 1988:153-154).

Chico Mendes worked as a rubber tapper in the town of Xapuri in Acre. His early life was typical:

> My life began just like that of all rubber tappers, as a virtual slave bound to the bidding of his master. I started work at nine years old, and like my father before me, instead of learning my ABC I learned how to extract latex from a rubber tree. From the last century until 1970, schools were forbidden on any rubber state in the Amazon. The rubber estate owner wouldn't allow it (Mendes, 1989:15).

In the early 1960's, however, he came across a political dissident from the Brazilian communist movement hiding in Acre. This man taught Mendes how to read and write, using political columns from newspapers. He also encouraged Chico to one day join union activities saying, "Look, you ought to get involved in trade union organization in this area. They will emerge, sooner or later, I don't know when, but this is where you ought to be. Don't avoid joining a union just because it is linked to the system, to the Ministry of Labour to the dictatorship" (Mendes, 1989:15). Mendes tried to organize the rubber tappers as early as 1968. The attempt, failed, however, since this was one of the most repressive periods of the military dictatorship.

The union would "arrive" through the *Confederação dos Trabalhadores na Agricultura* or the Confederation of Rural Worker's Unions in 1975. Acre then had no rural unions. The unions in Brasiléia and Xapuri were the first to be formed, but by the early 1980's rural workers' unions existed in all the municipalities of the state. Chico Mendes joined the union in Brasiléia in 1975. He was elected Secretary. Some of his friends from Xapuri joined this union and this led him to create a union there in 1977. Very early in his attempts he began to receive summons from the police. A catholic priest closely connected with the landowners and the government secret service told them about his activities (Mendes, 1989:20-21). In the words of Chico Mendes:

> . . . . The Xapuri union was founded with a great deal of self-sacrifice in April 1977. The local Church, the middle class and the local authorities put a lot of pressure on us, but despite this, the rubber tappers were very anxious to see things change and to be free from all the pressure and the threats. It all began quite slowly, but the question of organizing against the major cases of deforestation got under way again (Mendes, 1989:30).

In these years a preferred strategy used by the rubber tapper was the *empate*, or the standoff. In this strategy union people attempted to stop deforestation by blocking the path of machinery, people, and chainsaws, often with their own bodies. They would also lecture their opponents on the importance of the forest and why it was wrong to destroy it. Union members also became increasingly involved in the political process. Chico Mendes, was himself elected to the Xapuri municipal council in the late 1970's (Mendes, 1989:30).

While the *empate* and participation in local politics were important strategies of resistance to the devastation taking place in the state of Acre, they could only achieve limited success. The rubber tapper movement achieved political clout only after the creation of the National Council of Rubber Tappers in 1985, and when international interests in Amazônia reached high levels in the latter 1980's. It was only when the movement obtained international allies that it began to step on the Achilles' heel of the Brazilian government: the paving of BR-364.

As part of the POLONOROESTE Program, and responding to the increasing migration to the state of Acre, the government decided to extend and pave BR-364 from Rondônia into Acre, thus connecting the latter to the rest of Brazil by road. Rubber tappers like Chico Mendes feared that extending this road and paving it would further intensify the flood of migrants to their state. They were aware of the devastation that such roads could bring. When pavement of BR-364 through Rondônia was completed in 1984 it opened the gate to a flood of an estimated 160,000 migrants per year (Mahar, 1990:65; see also Martine, 1990). The result was the third highest rate of deforestation in Amazônia, an average rate of 2,340 Km^2 in 1978-1988-- roughly 1.11% (see Table 2).[20] Tappers in Acre were already under attack by powerful landowners and other economic interests who sought to "develop" their forests into cattle ranches.

Nearly half of the funding for the improvement of BR-364 was to come from the Inter-American Development Bank (IDB), an organization most-ly financed by the United States. With the help of the Environmental Defense Fund and the National Wildlife Federation, Chico Mendes was brought to the United States to lobby Bank officials against financing such a project. Also, American allies testified before the United States Senate's Foreign Operation Appropriation Sub-Committee on the ecological consequences of extending the road. When committee members warned the Bank that it faced the possible cutoff of U.S. funding, it withdrew its financial support. It was the first time in its history that the Bank stopped a loan on the basis of environmental concerns (Hyman, 1988:28; Schwartzman, 1991:411). The loan, then, was stopped by a combined effort of Chico Mendes' testimony and the political pressure of his international allies. And, Chico Mendes was fully aware of the importance of these allies:

> Our biggest assets are the international environmental lobby and the international press. I'm afraid we have more support from abroad than from the people in Brazil, as the opposite should be the case. It was only after international recognition and pressure that we started to get support from the rest of Brazil (Mendes, 1989:51).

This support and the results that it brought in terms of slowing road construction in Acre infuriated those who had the most to benefit from the project.   Chico Mendes and other union leaders became ripe for assassination.

We should point out that rubber tappers and Indians attempted to bridge differences by unifying their efforts. In 1986, the National Council of Rubber-Tappers and the Union of Indian Nations created the Amazonian Alliance of the People of the Forest in an effort to defend their ways of life. This allowed rubbers tappers to address the 3rd National Congress of the free trade union confederation, CUT, " ...calling on the Congress to defend the self-determination of the Indian nations and to strengthen the ties of solidarity between the forest peoples of Amazônia and workers all over Brazil, in order to prevent the fragmentation of struggles against projects such as Calha Norte..." (Treece, 1990:285). The larger aim of the movement was the demarcation of Indian lands and the creation of extractive reserves.

Reservations and Reserves

While there were differences in the aims of the Indian movement and that of the rubber tappers, e.g., constitutional recognition of Indian cultural differences, their movements converged on their desire to preserve the forests. For the Indians this means the preservation of their ancestral homes, and for the others the preservation of areas used for extraction. These protected areas would serve a function similar to U.S. national forests where nature is protected but some uses are allowed.

The creation of Indian reservations and extractive reserves has not come without controversy and conflict in Brazil. The demarcation of Indian reservations, as called by the New Constitution of 1988, has been especially controversial. For many high officials in Brazil, Indians are an obstacle to development and progress. For example, General Euclydes Figueiredo, a hardliner, argued that the Yanomami do not have the right to their traditional territory because "they are devoid of any intelligence, wandering around naked and breeding like animals" (in Weiss and Weiss, n.d.,p.8). The view espoused by many conservative nationalists is that too much land is being set aside for a handful of people who, by virtue of being "savages" do not know what is best for them, and can thus be easily manipulated by outsiders. In addition, as we saw, the military have traditionally viewed Indians with a great deal of suspicion, especially if they live near international borders or are divided by them, such as the division of the Yanomamis between Brazil and Venezuela (see *Veja*, August 25, 1993:26).

The reluctance to demarcate Indian lands is an extreme case of the common reluctance to set aside any protected lands in Brazil, such as ecological parks. In the dialectics of Amazônia, this reluctance reflects powerful systemic forces since access to the natural resources of these reservations would be restricted. The same applies to extractive reserves and to a lesser extent to national parks. In Indian reservations and extractive reserves, local groups become the "defenders" of the land; while national parks can be more easily plundered for lack of surveillance. When Indian lands are invaded complaints may become a source of international controversy. Even though invaders are not easily intimidated, the controversy can become sore spots in Brazil's international politics.

The Creation of Indian Reservations

Even though the New Constitution called for the demarcation of all Indian lands by 1993, the Sarney administration was slow in implementing the mandate. This was especially true of frontier areas which the military deemed important to national security. In areas under Calha Norte, for example, the military wanted Indigenous reservations substantially reduced (CEDI/PETI, 1990:7-9). Despite the reluctance there was a substantial increase in areas "officially recognized" as indigenous from 1987 to 1990, see Tables 3 and 4.

The data contained in Tables 3 and 4 suffer the ambiguity of Brazilian law. There are actually four categories applicable to Indian lands: 1) Non-Identified; 2) Identified; 3)Delimited; 4)Homologated; and 5) Regularized. Non-identified reservations are those in which no action to identify has been taken. Identification is done via an "administrative act" setting "physical boundaries" to lands which belong to the Indians; it amounts to recognition of a claim of existence. Delimitation means that a document has been issued by the government recognizing the existence of the reservation, on paper. Demarcation is done by topographically studying the area and placing physical markers to establish reservation boundaries. Homologation means that the reservation is recognized by the Federal Government, by presidential decree. Finally, regulating the reservation means that the proper documents are filed with the government, in the past with the Serviço do Patrimônio da União (SPU). These categories are not always clear cut. For example, the delimitation phase has often been done by different officials: FUNAI's president, State Ministers, or the Brazilian President. The documentation may thus vary from case to case (CEDI/Museu Nacional, 1987:11).

Despite the vagueness of the categories, they are often grouped into two

Table 3. *Recognition of Indian Lands by the Brazilian State, 1987*

State of Recognition	Number	%	Area	%	Population	%
Non-Identified	167	32.24	0	0	10,245	4.80
Identified	107	20.66	37,520.703	50.39	67,290	31.54
Delimited	171	33.01	32,117.459	43.13	96,505	45.23
Homologada	32	6.18	1,949.628	2.60	18,036	8.45
Regularized	41	7.91	2,887.359	3.88	21,276	9.98
Total	518	100.0	74,466.149	100.00	213,352	100.00

Source: CEDI/Museu Nacional, *Terras Indígenas no Brasil* (Rio de Janeiro: Museu Nacional 1987)

Table 4. *Recognition of Indian Lands, 1987 and 1990*

Land Category	Year	Number	Area (ha)	Population
Officially Recognized	1987	351	74,466.149	203,107
	1990	436	79,097,854	228,814
Non-Recognized	1987	167		10,245
	1990	90		6,802
Total	1987	518	74,466.149	213,352
	1990	526	79,097.854	235,616

Source: CEDI/PETI, Terras Indígenas no Brasil (Rio de Janeiro: Museu Nacional 1990)

categories: Officially Recognized, and Non-Recognized. In 1987 there were 351 Officially Recognized reservations encompassing an area of 74,466,149 hectares and including a population of 203,107 Indians. By 1990 the number increased to 436 reservations occupying 79,097,854 hectares with a population of 228,814 (see Table 4). However, the organization that reports these data, Centro Ecumênico de Documentação e Informação (CEDI) or Ecumenical Center for Documentation and Information, cautions that the data can be misleading. Indian lands are not "physical objects that can be easily defined in form and quantity, thus allowing for simple arithmetic" (CEDI/PETI, 1990:6). Lands are defined, but reality is something else. Of the 167 reservations listed in Table 4 as Non-Recognized in 1987, only 38 moved into the Recognized category by 1990. Also, "recognition" does not mean "homologation," meaning that the status of the reservation remains vulnerable. By the beginnings of the Collor administration 237 reservations out of 526, 45% percent, were not yet recognized (CEDI/PETI, 1990:9).

Actually, the Collor administration was more receptive to the demarcation of Indian reservations. Both the Kayapó and the Yanomami reservations were established during this administration. The Kayapó reservation comprises an area of 4.9 million hectares, slightly larger than Switzerland, and the Yanomami 9.4 million hectares, roughly the size of Portugal. Collor also established smaller reservation such as the 35,922-hectare Caigangue Indian reservation in the southern state of Rio Grande do Sul. On the occasion of legalizing this reservation he stated, "To protect the Indian is the patriotic public spirit of a government official." He cited the American film *Dances With Wolves*, adding that he did not want the same destruction of Indian culture to happen in Brazil (*Jornal do Brasil*, March 28, 1991). In a very brief period of time, Collor protected half again as much Indian land as was set aside in the preceding 80 years (Inter-American Development Bank, United Nations Development Programme, and Amazon Cooperation Treaty, 1992).

Extractive Reserves

The struggle by the rubber tappers to have their reserves established was equally difficult. According to Mary Helena Allegretti (1994:22-24) the history leading to the creation of the reserves can be divided into four periods: 1) *Empates* and Expulsions (1973-1976); 2) Agreements and Compensation (1976-1980); 3) Colonization (1980-1985); and finally 4) Extractive Reserves (1985-1990). In the *Empates* and Expulsions period the rubber tappers used the *empate* as a strategy against expulsion. They simply walked to the areas being destroyed and sabotaged the camping sites before operations began. They also filed law suit against the new landowners. By 1988 more than 40 *empates* had taken place throughout the region. The Agreements and Compensation period was characterized by recognition by the federal government of the rubber tappers' de-facto possession of the land. This meant that rubber tappers had to be compensated before being expelled from the land, a legal fiction in a land where guns talk. In the Colonization period, the federal government attempted to intervene in the conflict by creating colonization areas to which rubber tappers, and small producers, would be relocated. The aim of these projects was to transform rubber tappers into agriculturalists. This failed because of lack of an agricultural tradition among rubber tappers, and the lack of infrastructure in the region. Many of the small producers in these colonization schemes actually converted to extractive activities, rather than rubber tappers converting to agriculture. The market for extractive goods was more stable than in agriculture. Finally, at a national meeting of Amazônia rubber tappers in 1985 it was decided that the main objectives of the movement would be for rubber tappers to remain in the forest and to demand agrarian reform that would respect this traditional way of life. The movement also aimed at obtaining modern production technologies, and the creation of education and health systems adapted to the needs of the community. It was only after the death of Chico Mendes that the 970,570-hectares Chico Mendes Extractive Reserve, accommodating 3,000 families, was created (Allegretti, 1994:29-30; Assis Menezes, 1994:70).

The very idea of extractive reserves was thought incongruent with the legislation of the time, since this defined property in individual terms rather than collectively. Decree No. 98.897/90 was the compromise. It defined these areas as "property of the union or federal government," destined to be used for extractive activities. The extractive population would have de facto access granted to it but the reserves were property of the state (Allegretti, 1994:29-30; Geisler and Silberling, 1992:58).

In addition to the creation of extractive reserves in areas already populated by rubber tappers the government decided to create Extractive Relocation Projects as an alternative to the traditional agricultural schemes. In 1989 eleven such projects were already established in the region with an area of 889,548 hectares and benefitting 2,924 families. In the same year four extractive reserves were established in Amazônia comprising an area of 2,162,989 hectares and benefitting 6,250 families(Assis Menezes, 1994: 69-70).[21] By 1994, two more extractive reserves were created bringing the total number to nine and increasing the extractive-reserve area to 2,200,750 hectares (Murrieta and Rueda, 1995:9). Twenty years earlier the rubber tappers had nothing.

The progressive nature of extractive reserves provides an alternative, environmentally sustainable model of incorporation for the region. They have attracted the support of international organizations such as the Worldwide Fund for Nature, the Ford Foundation, and the Conservation Foundation in addition to several Brazilian organizations. They should be viewed as "experimental models" that will likely be utilized in other parts of the world if successful (Geisler and Silberling, 1992:59-60).

Demarcation: An Analysis

The demarcation of Indian reservations and extractive reserves was a step in the right direction, but the picture looks better on paper than in reality. The demarcation was achieved against a background of international pressure. Without this pressure the process may not have occurred nor does the creation of these reservations mean that Indian lands and extractive reserves are being protected. Both the Yanomami and the Kayapó reservations occupy huge areas and thus necessitate a large personnel to monitor them. Even though the Collor administration took steps to remove gold miners from Yanomami lands, for example, it did not provide means to monitor their return. Yanomami lands have continued to be invaded by gold prospectors. In 1996 there were thirty-five clandestine airstrips and an estimated 3,000 gold miners inside Yanomami territory (Rainforest Action Network, 1996).

Again, the creation of reserves should be viewed against a background of global ecopolitics. Democracy allowed local groups to organize and make demands. But, they were successful in converting those demands into victories only when global ecopolitics intervened in their favor. It also should be kept in mind that this international pressure came about as a result of a perceived necessity. As people began to fear global climatic change, they finally turned their attention to the fate of tropical rainforests and their inhabitants. These were ignored earlier.

For Every Action a Reaction

The growing international attention received by grassroots organizations in the 1980's annoyed Brazilian conservatives. They viewed these movements as anti-Brazilian, and saw the speeches given abroad as an attempt to demoralize Brazil. So the Indian movement became a special target during the writing of the New Constitution. The anti-Indian lobby was spearheaded by the conservative newspaper *O Estado de São Paulo*.

In 1987 this newspaper published a series of articles entitled "*Os Índios na Nova Constituição*" or "The Indian in the New Constitution" which claimed there was an international plot by CIMI and foreigners against Brazil:

> As it will be seen throughout this series the campaign in favor of preserving Indian lands of any intrusion by the "white man" was planned abroad and financed by money deposited abroad since 1981. The documents we are going to use are very clear: they are texts containing specific guidelines for Brazil, letters of higherups in the CIMI hierarchy, statement by CIMI's president...and bulletins from that organization. From these a sad picture emerges of men of good faith involved in dirty maneuvering, men used by the interests of international groups-- **which are not interested in Brazil. On the contrary, they do not want investments in the [Brazilian] mineral sector to occur in order for them to maintain their monopolies in the international markets** (*Estado de S. Paulo*, August 9, 1987, p.4).[22]

The newspaper accused CIMI and the Catholic Church of "conspiring against Brazil." It claimed that the Church was involved in a deliberate plot to restrict Brazil's sovereignty in Amazônia. It based its assumptions on a document, "Diretrizes Brasil No.4-Ano 0" or "Brazil Guidelines No.4-Year 0" issued by the International Council of Christian Churches in a meeting held in 1981 in Genebra. The newspaper's interpretation of the document was that it provided specific guidelines for the international-ization of Amazônia. For example:

> The Guidelines No.4 geared towards Brazil begins with a declaration: "The totality of Amazônia, most of which falls within Brazil, also encompasses parts of Venezuelan, Colombian, and Peruvian territories. It is considered by us a property of humanity. **The possession of this immense area by the mentioned countries is merely circumstantial** (emphasis added by the editors), this was not only decided by the organizations present in this symposium, but it is also the philosophical decision by the more than one thousands members that constitute the several councils for the defense of Indians and the environment. . . ." From this declaration of principles normative instructions follow which begin always with "It is our duty . . . ." The first one established that all resources must be used to preserve this immense territory [Amazônia] which is "property of humanity and not the property of the countries whose territories they pretentiously claim as their own" (*O Estado de S. Paulo*, August 9, 1987, p.4).[23]

The rationale for publishing this series of articles was a petition letter sent to the writers of the New Constitution by an Austrian Catholic youth group in support of Indian rights. For the newspaper this group had ulterior motives. It wrote:

> The campaign organized by the Dreikoenigsaktion der Katholischen Jungschar Oesterreich (Austrian Catholic Youth Group) attempting to influence the decisions of the Constitutional National Assembly on Indigenous problems shows the validity of the fears expressed by the Secretariat of the Council of National Security in 1985, "of national and international pressures aiming at building--at the cost of the current Brazilian and Venezuelan national territories--an Yanomami state." If the objective of the campaign is not the creation of a Yanomami state, then it aims at subtracting 14% of legal Amazônia from the strict control of Brazilians via the implementation of the concept of "restricted sovereignty" in indigenous areas. This will be accomplished by altering the current legislation....(*O Estado de S. Paulo*, August 9, 1987, p.4).[24]

Incredible as these accusations now seem, President José Sarney called for a full investigation by the Brazilian Institute of Investigation (SNI). The accusations also gave the military an opportunity to ventilate its dis-pleasure. They condemned the idea of "restricted sovereignty" over Indian lands in Amazônia as a foreign plot to takeover the region. A military minister, General Paulo Campos Paiva, reminded the press that in 1904 the Minister of War, Marshal Francisco de Paulo Argollo, argued that there was international covetousness which could lead to an invasion of the region. Campos Paiva also stated, "... there is an international conspiracy and we have to defend ourselves against false missionaries . . . ." One of the stories in the same series was titled *"Tropas estão prontas para o que der e vier,"* "The Troops are Ready for Whatever Comes," suggesting an imminent foreign invasion of Amazônia (*O Estado de S. Paulo*, August 12, 1987, p. 5).[25]

The Brazilian National Council of Bishops (CNBB) reacted by threatening to sue the newspaper. According to the Council, the series "showed a malicious interpretation of the activities of the Church and it was based on totally false claims." CIMI's Executive Secretary, Antônio Branndt, also argued that the articles "[revealed] a power lobby which [aimed] at sweeping once and for all any possibility of victory [in the *Constituintes*] by the Brazilian Indians" (*O Estado de S. Paulo*, August 11, 1987, pp. 4-5).[26]

The charges against CIMI came at a cost to the Indians. The constitutional clause declaring Brazil a multi-ethnic society was voted down (Maybury-Lewis, 1989:3). In the end, most of the legal provisions fought for were approved; perhaps it became clear to the politicians that these were legitimate claims, not a plot to allow foreign powers to take over the country. It was also clear that Indians and their allies would not simply be pushed away. This time they were organized and their lobby was strong. Their efforts were orchestrated into explosive media events. When, for example, the Indians delivered their constitutional demands to the writers of the New Constitution they came in full traditional attire and in groups, making headlines in Brazilian newspapers (see *O Estado de S. Paulo*, August 13, 1987). In an emerging democracy threats and accusations were no longer enough to make people retreat. This time people could defend themselves in public, in view of the Brazilian people.

The alarmist charges often backfired against the conservatives. This was certainly the case of charges against Kayapó leaders Paulinho Payakan, Kube-i and American ethnobotanist Daryl Posey, who traveled to the United States in late 1988. They spoke against Brazil's massive hydro-electric plans for Amazônia at a conference at the University of Southern Florida and then proceeded to Washington, D.C., to meet with World Bank leaders. Upon their return to Brazil, they were charged with "defaming the national image," a crime punishable with up to three years in jail.[27]

The Kayapós used these accusations to advantage. On the day of the hearing the accused showed up accompanied by 200 Kayapó warriors armed with clubs and bows and arrows. They harassed the military police in the surrounding courthouse while the judge and his assistants barricaded themselves inside. This was a major media event:

> The local press was mesmerized. Photojournalists and televisions crews fell over themselves attempting to capture these extraordinary images. During breaks between ceremonial dances and ritualized berating of the military police, Payakan placed a map on an easel in the middle of the street and gave impromptu lectures to the tribesmen on the impact of the dam scheme and his motives for appearing before the World Bank. . . . [Once the Altamira event started], the charges mysteriously disappeared in the Brazilian courts (O'Connor, 1997:122-123).

This type of mass demonstration was not new to the Kayapós. In 1987, when the Indian issue was at risk during the constitutional convention, pro-Indian activists mounted a campaign to save what they could. The campaign was aided by "...a contingent of Indians, mostly Kayapó, who sat, painted and feathered, day in and day out, inside the parliament buildings, impressing the framers of the new constitution with their fortitude and persistence" (Maybury-Lewis, 1989:4). As we saw, the multi-ethnic clause was lost but major victories were won.

The failure to prosecute Payakan, Kube-i, and Posey shows that the nationalism inherited from the dictatorship would no longer stop people from claiming their rights. The charges became controversial in the eyes of the outside world. The government no longer had carte blanche in its dealings with the Indians. But, even success has its price.

In the dialectics between the topdogs and underdogs of Brazil, actions lead to reactions in an endless political cycle. In 1985, landowners formed the *União Democrática Ruralista* or the Rural Democratic Union (UDR). Their purpose was to influence the writing of the New Constitution. Their financial power is illustrated by the fact that during a political rally the airport in Brasília ran out of parking spaces because of hundreds of private airplanes (Mendes, 1989:58). On a local level, UDR meant the political intimidation of rural workers through increased violence. As Chico Mendes' movement increased in importance it became the target of more violence and intimidation.

> The other tactic the landowners use, and it's a very effective one, is to use hired guns to intimidate us. Our movement's leaders, not just myself but quite a few others as well, have been threatened a lot this year. We are all on the death list of the UDR's assassination squads. Here in Xapuri, these squads are led by Darli and Alvarino Alves da Silva, owners of the Parana and other ranches round here. They lead a gang of about 30 gunman--I say

30 because we counted them as they patrol the town (Mendes, 1989:66).

The two mentioned by Mendes in the above quote, Darli and Alvarino Alves da Silva, eventually killed him.

> It was in this process [union and grassroots organizing] that leaders such as Wilson Pinheiro, Jesus Andre Matias, and Chico Mendes became known. It is not by accident that all of these leaders have been assassinated--to the extent that the movement succeeded in interfering in land holding relations, large landowners reacted by attempting to eliminate the leadership (Schwartzman, 1991:404).

Reaction against the Indians also included not only invasion but murder. The demarcation of Indian lands in the Northeast of Brazil has led to increased conflict between Indian and farmers in the region. When Xukuru Kariri Indians invaded two farms located in their lands on September 8, 1994, farmers retaliated by threatening to kill them and their missionary friends (CIMI, 1994b). Patoxó Hã-Hã-Hãe Indians were faced with a similar conflict with farmers when their reservation was demarcated (*Jornal do Brasil on Line*, May 2, 1997). The demarcation of Yanomami and Kayapó lands has also led to intense struggles with gold miners who keep returning to these reservations after being expelled (see CIMI, 1992b and 1994c).

Statistics on violence against Indians have been compiled by CIMI. For 1990, 1991, and 1992 the organization reports 13; 26; and 42 murders, respectively. These figures are aggregates, however. The 42 murders in 1993 were committed against Indians by both Indians (13) and non-Indians (29). The reasons behind them are not understood, due to the poor quality of police investigations as evidenced by the fact that though "most cases were officially investigated . . . the offenders were not punished." "Because of the prevailing impunity, the violence against Indians tends to grow" (CIMI, 1993b).

CIMI's statistics for violence against Indians for 1992 is more detailed. It reveals the extent of the violence (see Table 5). Even though correlation does not mean causation, it appears that violence against Indians is on the rise, thanks to land disputes.

One recent case has become the symbol in the fight for justice against those who perpetrate violence against Indians. On April 20, 1997, a Patoxó Indian, Galdino, was burned alive by a group of teenagers while sleeping at a bus stop in Brasília. Galdino went to Brasília to participate in the celebration of the Day of the Indian and to march with the landless.

Table 5. *Violence Against Indians, 1992*

Category	Number of Cases
Murders	24
Victims of Attempted Murder	20
Death Threats	21
Illegal Detentions	5
Physical Aggression Victims	10
Rapes	7
Suicides	24
Malaria Deaths	87
Measles Deaths	64
Cholera Deaths	14
Judicial Decisions Against Indian Interest	5
Car Accident Deaths	6
Areas Invade by Woodcutters	37
Invasion by Miners	16

Source: CIMI, "In 1992. 24 Indians Were Murdered in Brazil," in "Indian and Violence in 1992" (03/93), Brazilian Conservation Documents Prior to 1977, Gaia Forest Conservation Archives <http://forests.org/>

When he returned to the boarding house where he was staying the doors were locked, so he slept at the bus stop. After spotting Galdino the teenagers went to a gas station and bought two liters of alcohol with which they set Galdino on fire. But the crime had witnesses and the teenagers were caught. One was the son of a top judge in Brasília. They stated that they had no intention of killing Galdino, it was all a prank. What led to public indignation, was that after a hearing the charges were reduced from homicide to unintentional homicide. According to Brazilian law, this lesser punishment meant that the trial need not go to jury. The judge's verdict was that "Burns do not kill." He had, he said, seen people survive burns. The implications for Indigenous people were summarized by Dom Pedro Casaldaliga, Bishop of Prelazia de São Felix do Araguaia in the state of Mato Grosso:

> This is simply another moral suicide for Brazilian justice. It is one more legal inequality that has hit Indigenous people. It has been demonstrated that justice is only valid when it is applied to the poor. I am outraged and I hope to see from those who do not wish to see the continuation of the Indigenous problem an ethical and public outcry against the decision (in *Jornal do Brasil*, March 6, 1998).[28]

After UNCED

After the hoopla created by UNCED and the Global Forum, interest in the environment declined. There were reports about the state of the environment in the international media, but they did not stir the media's interest in Brazil. This lack of interest by the popular media undermined gains made in the prior decade. Itamar Franco's administration did very little to further the gains. However, the Cardoso administration, being the farthest in time from UNCED, succumbed to pressure from the conservatives and attempted to reverse things. On January 8, 1996 the President signed a decree that allowed already demarcated Indian lands to be challenged in court by private individuals and corporations. With a single stroke of the pen he reopened legal challenges against Indian reservations. However, this move was a gross misreading of the international situation. Even though the media attention on Amazônia had dwindled, there remained a significant number of interest groups, and they mobilized a "virtual" campaign on the Worldwide Web against the government of Brazil and its decree, along with demonstrations in the "real" world. They put pressure on the G-7 countries. This led the European Parliament to pass a resolution on February 15 of the same year condemning the decree. This was worrisome:

> Worried with Brazil's image abroad, president Fernando Henrique Cardoso quickly summoned the president of FUNAI, Marcio Santilli, to the Planalto Palace for a meeting on Saturday afternoon. The results of the meeting have not been disseminated, but information got about that Santilli was instructed to remain "on alert." The Brazilian government decided not to cancel the trip of minister Nelson Jobim to Europe, which is scheduled to May 25. Arrogantly, Jobim says that Brazil will not be forced to provide explanations

to the international public opinion and that the aim of the trip is to present the National Human Rights Plan. . . (CIMI, 1996b).

Despite his original arrogant tone, when in Europe the minister attempted damage control:

> Worried with the negative repercussions of the decree, minister Jobim met, also on Wednesday, the ambassadors of the G-7 countries, Nordic countries and of the Vatican to explain why the measure was taken. Speaking for almost two hours, Jobim defended the adversary system in meetings. The ambassadors took no position in relation to the issue (CIMI, 1996c).

Despite the original reaction by NGO's the decree was less devastating than feared originally. The Rapôsa Serra do Sol reservation in the state of Roraima was reduced by 300,000 hectares, and the demarcation of other reservations was delayed (see Environmental News Network, 1997).

Indians and Caboclos Become Lumber Jacks

Perhaps a major future challenge to the movement for the people of the forest will result from their increasing involvement in capitalist activities in the region. As Indians and *cablocos* turn to economic activities that are not considered "environmentally correct" by their international environmentalist allies they may lose political support from NGO's. As we pointed out above, when the Kayapós signed a contract with a lumbering company, the rock star Sting withdrew his support. The international environmentalist movement still has a late-20th-century perception of Indians as "noble savages," as protectors of the rainforest. As the case of Colonel Tuto Pompo indicates, Indians are also growing aware that in the world economy that is engulfing them, "economic resources" are imperative for survival. Thus, under his leadership the Kayapós involved themselves in gold and lumbering activities. But, money corrupts. There have been charges, mostly by the Brazilian press, that the chiefs often take advantage of their positions to extract a disproportionate share of the returns. The Brazilian magazine *Isto É* (Oliveira, 1993), said that Colonel Pombo amassed a fortune worth six million U.S. dollars, which included two airplanes and three farms. His successor, Tapiet, speaks fluent Portuguese and has also amassed wealth. He has two houses, 100 head of cattle, a 346 alqueire[29] farm, two airplanes and several cars, all obtained by taxing gold prospectors operating in Kayapó land. Tapiet denied access to the Santillo mine to people not involved with the mining operations. According to the magazine, this was to avert criticism from environmentalists:

> But what is the secret about Santillo that it cannot be seen? "There is
> nothing to be seen there. The ecologists keep on criticizing us," says Tapiet.
> In addition to having opened a crater in the jungle and having contributed to
> the pollution of the river Fresco, already decimated by mercury by the Maria
> Bonita Mines, Santillo certainly does not cause any additional harm. Only
> to the image of ecological purity of the Kayapós, an image they try to
> maintain at all costs. Purity seems to be a quality Tapiet lost a long time ago
> (Oliveira, 1993, p.60).[30]

By becoming capitalists and by excluding environmentalists, the Kayapó
run the risk of losing their legitimacy and their allies.

It has also become evident that unsustainable lumbering is a major
economic activity in Amazônia, making Brazil the third largest producer
of tropical woods in the world in 1998. Much of the harvest of trees is
done illegally by *"cupins*, "termites," or poor individuals such as *caboclos*
and Indians who sell trees, such as mahogany, at banana prices to
middlemen, who in turn sell them to multinationals. Mahogany is bought
from the *cupins* for less than US$30 dollars (R$30) and sold for close to
US$800 dollars (R$800) in the market (*O Globo*, March 8, 1998). While
this is done mostly by *caboclos*, Indian involvement in logging is growing,
sometimes with the help of Brazilian NGO's who see this as a way for the
Indians to maintain themselves. For example, the Xikrin Indians of the
Carajás area in the state of Pará have been trained by the Brazilian NGO
Instituto Socio-Ambiental (ISA) to patrol the selective cutting of
hardwoods from their reservation by the Companhia do Vale do Rio Doce.
The World Bank has contributed US$400,000 to the project (Environ-
mental News Network, 1998). Thus, direct Indian involvement in the
lumber industry will probably continue to increase. According to *O Globo*
(March 8, 1998):

> Pressured on one side and co-opted on the other, the Indians are giving in to
> the white man and faced with their invasion are participating directly in the
> lumber business. . . . FUNAI has admitted that at least 60 different Indian
> lands, the majority in Amazônia, are being persistently encroached upon by
> lumber companies.[31]

The same newspaper quotes Xikrin chief Karangre as stating that "Things
will work out all right. The sale of wood will be made in an organized
fashion. It is good for Brazil and for the Indians" (March 8, 1998, p.15).[32]
It is questionable that lumbering can actually be done "in organized

fashion." For some Brazilian environmentalists the wealth contained in the trees of Amazônia is so great that people will continue to cut beyond established limits. Some have proposed a ten-year moratorium in lumbering contracts and a ban on the cutting of mahogany (*O Globo*, March 10, 1998). How such a moratorium will be enforced has not been addressed. Would it be another law that looks good on paper but lacks enforcement? If Indians allow environmentally unsound practices in reservations that environmentalists thought would be preserved, what will be the consequences to their movement for self-determination? Will environmentalists help only if the resources in the reservation are not sold? Will environmentalists have to choose between Indians and trees? As Indians adopt capitalist rationality will they be perceived as rational or irrational?

Notes

[1] *Caboclo* is the general term for the members of the rural lower class in Amazônia. The term is often applied to people of "Indian-"white" ancestry who live in isolated areas along rivers (Fearnside, 1986:232)

[2] For a description of the situation of the Urueu-Wau-Waus see von Puttkamer (1988:814).

[3] See for example José de Alencar's *O Guaraní* (n.d.[1857]) .

[4] Author's translation.

[5] A major problem for Brazilian prosecutors in this case was the Yanomami practice of endocannibalism. Endo-cannibalism among the Yanomami consists of the practice of cremating the dead, liquefying the ashes and drinking them. It is done so that the spirits of the dead have a place to stay. For Brazilian law officials this meant that there were no bodies, thus no evidence (see O'Connor, 1997:357-59).

[6] Author's translation.

[7] Author's translation.

[8] Author's translation.

[9] Author's translation.

[10] Author's translation.

[11] Author's translation.

[12] Author's translation.

[13] The figures for population were obtained from the Citations. The same document refers to 13 Kayapó villages, rather than the 14 referred by Fisher (1994).

[14.] This was part of the Kayapó's attempt to clear the western portion of their reservation from Brazilian intrusion. It should be kept in mind that the demarcation of their reservation had been interrupted at this stage in time since 1978 (da Silva et al., 1985:120).

[15] Author's translation.

[16.] Author's translation.

[17.] Author's translation from the Portuguese translation.

[18.] Author's translation.

[19.] It should be kept in mind, however, that there is a fundamental difference between the two groups. The rubber-tappers have been fully integrated into a capitalist world-economy. They depend on the sale of latex for their livelihood. Indians have also participated in this economy, e.g., by selling Brazilian nuts, however to a much lesser extent. Regardless of these differences, however, both groups had a major commonality: their dependence on the blessings of the forest.

[20.] This rate was only smaller to Maranhão's 1.79% and Tocantins' 2.97% in the same period, 1979-1989 (INPE, 1998, 1999).

[21.] Date for extractive reserves in this period varies slightly from author to author. Philip Feanside (1989:389) reports 20 extractive reserves "Already Existing" or "Proposed," comprising an area of 22,161 km^2 (2,216,100 hectares) and being home to 2,290 families. He also lists reserves not included in Assis Meneze's data. Stephen Schwartzman (1991:418) provides an even higher estimate of three million hectares occupied by 14 reserves, the Chico Mendes reserve being the biggest.

[22.] Author's translation.

[23.] Author's translation.

[24.] The Austrian group obtained 47 thousand signatures (*O Estado de S. Paulo*, August 11, 1987).

[25.] A harsh statement was also made by the head of the Federal Police, Romeu Tuma, who argued that the government should analyze the situation from the perspective of wealth. For him, Amazônia is a region containing mineral deposits of strategic importance, deposits vital for the national economy. According to him, the presence of these minerals lead to envy by countries which have already exhausted their natural resources. He pointed out this is the main reason the government was implementing the Calha Norte project, as a way of promoting the occupation of the region (*O Estado de S. Paulo*, August 12, 1987, p.5).

[26.] Author's translation.

[27.] This strategy of intimidation continues to be used. On March 17, 1998 a Dutch missionary working for CIMI had his visa duration reduced as a result of an interview he gave about the Tupiniqui and Guarani's demonstration for the demarcation of their reservations. These Indians have been involved in a struggle since they, with the help of the missionaries, decided to demarcate the reservations themselves. The missionary was indited under the Foreigners Law which restricts the amount of political activity foreigners can engage in Brazil (*Jornal do Brasil*, March 19, 1998).

[28.] Author's translation .

[29.] Alqueire is a Brazilian scale for area. It varies in different parts of the country. In Rio de Janeiro, Minas Gerais, and Goiás it is worth 4.84 hectares. In the state of São Paulo it is worth 2.42 hectares.

[30.] Author's translation.

[31.] Author's translation.

[32.] Author's translation.

Chapter 7

Comparisons With Costa Rica, Malaysia, and Indonesia

It is important to start this chapter by restating some of our original claims on democracy from Chapter 1. First, we see democracy as a facilitator of environmental protection. Relative to authoritarianism, it is conducive to debates about the state of the environment, thus having the potential to create solutions. There is no perfect correlation between democracy and the environment, however. Environmental degradation can occur under democracy, since democracy alone is no guarantee of preservation. However, we argue that for true solutions for the problem of environmental destruction to emerge, public debate is necessary, thus the presence of democracy is a condition sine qua non for this to occur. But before democracy can play a role in protecting the environment, society must attach a certain degree of importance to the preservation of nature; its culture has to value it. For societies which have lacked this tradition, democracy in the late twentieth century has become an important factor for the diffusion of preservationism from the outside world. As a political system it makes countries porous, more subjected to the dynamics of global culture. As the ecopolitics of the world-system greens, and its rhetoric becomes greener, democracy allows for environmental concerns and appreciation to be brought in, or to be resurrected in people's minds. Authoritarianism is the antithesis of all of this. In developing societies characterized by authoritarianism, ideas such as "nature as source of raw materials," as "jungle which is in the way of development," "deforestation for development," "development at any cost," etc., go unchallenged because of repression.

It is important to point out that we are not oblivious to the fact that economic democracy is as important as political participation for the preservation of the environment.[1]   Poverty breeds environmental destruction. For poor people, tropical rainforests often remain their only alternative, their only possibility of getting a plot of land on which to grow food. Democracy does not necessarily lead to economic equality as evidenced by countries such as India where tremendous inequalities continue to exist. It is our contention, however, that economic equality alone is not enough for environmental preservation to occur. A political culture of preservationism must be present; and, where it does not exist it will not emerge on a national level without democracy, regardless of economic equality. Concerns about environmental degradation such as pollution and water quality emerge in non-democratic regimes, but these concerns fail to be translated into a national culture of preservation, or even national policy. For government leaders anxious to achieve economic development the emergence of an environmentalist movement is often viewed as dangerous.   For them, preservation is the antithesis of development, an enemy to be avoided.

In this chapter we compare Brazil to the cases of Costa Rica, Malaysia, and Indonesia as a way of demonstrating the soundness of the above arguments. We chose these societies because they are at opposite ends of an imperfect political continuum, Costa Rica is at one end as the most democratic and Malaysia and Indonesia at the other as the least. Since the early 1990's, Brazil has moved closer to Costa Rica in political maturity. These societies were also chosen because, like Brazil, their efforts towards economic develop are deemed high; that is, their degree of participation in the capitalist world-economy is high.   They are also societies where systemic forces of destruction are strong, where there is a strong impetus to use natural resources for the sake of development, and this has certainly reflected on their deforestation rates.

What we are most concerned about is the impact of democratic freedoms on what we call the "Preservation Effort,"[2] which we define as the willingness to set aside forest land to remain undisturbed.   The preservation effort for us is different, however, from forest management for commercial use, which involves the destruction of the original forest. We use both quantitative and qualitative information for each country as indicators of this effort.   We use two quantitative indicators: 1) Deforestation; and 2) Natural Reserves. Deforestation is defined here as area cleared per annum (km^2/annum) as well as the decline in the total forest area of each country. The definition of Natural Reserves is how much of the total forest area is considered legally protected. It should be noted that we treat Indian reservations and extractive reserves as protected, even though we indicate those figures separately. Qualitative information such as statements and situational accounts are used to balance the quantitative evidence. This is necessary due to the conflicting nature of most quantitative evidence. For example, measures of deforestation are not without controversy. Until recently, most measures were provided by a combination of methods such as aerial photographs and agricultural surveys. Thus, estimates tend to vary somewhat depending upon the method and the source (Brookfield and Byron, 1990:46). The data for some countries have improved substantially in recent years due to the use of satellite technology. Brazil's INPE, for example, has relied substantially on this technology to generate new and revise old data.   Similar controversies surround data on forest cover and protected areas. Some countries include areas of forest management for the timber industry as part of their permanent forest cover. Also, legal protection does not mean de-facto protection, and some protected areas in countries such as Indonesia and Malaysia have reverted back to logging without public consideration.   Despite these pitfalls, however, we believe that a combination of quantitative and qualitative information can provide us with a fair picture of events.

The arguments presented above are summarized in Figure 1. Figure 1 puts Brazil, Costa Rica, Malaysia and Indonesia along three continua: 1) Democracy vs. Authoritarianism; 2) Development Effort; and 3) Preservation Effort. It suggests that democracy is an intervening variable, attenuating the developmentalist drive of governments in favor of the environment. It should be noted that we grouped the four countries on either side of the continua, in general categories, rather than specifically ranking each country. The understanding of how democracy impacts the environment is our objective, not the ranking of countries. Ranking is only used throughout the text when it helps achieve our objective. We begin with the case of Costa Rica.

COSTA RICA

Costa Rica is an excellent case to further demonstrate the arguments put forth in this book about democracy as facilitator of environmental preservation in the evolving market economies. Like Brazil, Costa Rica's problems of deforestation have been a result of an expanding agricultural

Figure 1. *Placement of Cases Along Three Continua*

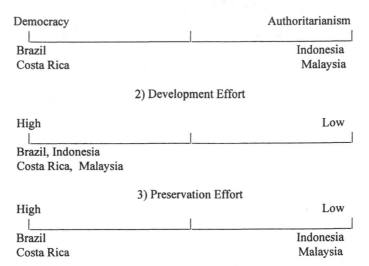

1)Democracy vs. Authoritarianism

Democracy   Authoritarianism
|_____|_____|
Brazil   Indonesia
Costa Rica   Malaysia

2) Development Effort

High   Low
|_____|_____|
Brazil, Indonesia
Costa Rica, Malaysia

3) Preservation Effort
High   Low
|_____|_____|
Brazil   Indonesia
Costa Rica   Malaysia

frontier.   Forest lands were also viewed in that country as "free for the taking," their occupation defined as a form of improvement or development. As in Brazil, "a strong constituency for deforestation exists in Costa Rica, including small farmers, large farmers, agro-exporters, bankers, and urban consumers, all of whom pressure the government for policies which ultimately result in deforestation" (Jones, 1992:681). That is, despite differences in size and geography both countries have been under pressure to develop their rainforests, to open them up for settlement and production. As indicated above, this development pressure is central to our arguments on democracy. The point is that pressures to expand the frontier can be attenuated only through a democratic process that includes all of those involved. One might find vast forests intact under authoritarian regimes if a country is economically stagnant. Once development begins to occur, these forests are quickly utilized for economic purposes without a public forum, however. As we shall see, to a large extent this explains the devastation of rainforests in Asian countries such as Indonesia and Malaysia.

A major difference between Brazil and Costa Rica is that the latter has the oldest and most stable democracy in Latin America, while the former's recent experiment with that system is 14 years old (taking civilian rule in 1985 as the beginning). In terms of democracy, Costa Rica is at the forefront in Latin America. The fact that it is the oldest democracy in the region to a large extent explains the environmental edge it has over other countries. But, just as in the Brazilian case, the relationship between democracy and environmental protection has to be viewed within the context of changes in the ecopolitics of the world-system. Concerns about preservation in Costa Rica intensified in the mid-1980's also, when a culture of environmentalism infiltrated it from abroad. This is not to say that preservation was not an issue before. The history of environmental legislation in that country attests to this.

Legal requirements for controlling deforestation in Costa Rica have existed since 1841, when laws requiring the replanting of areas proportional to cleared ones were passed. These requirements aimed at maintaining a supply of medicinal and construction grade species. In 1942 a "Ley de Aguas" (Law of Waters), designed to protect water course and water supplies was also passed. However, these laws were to a large extent ignored or were difficult to enforce. From the 1960's through the 1980's new laws were passed to protect the environment, e.g., the Forestry Law of 1969 and the Reforestation Law of 1976, but these suffered poor enforcement (Jones, 1992:684-685). Despite the lack of enforcement most of Costa Rica remained under forest cover until the 1950's. As late as 1970 more than half of its territory remained forested. The 1970's, however, changed this situation drastically. Deforestation rates began to rise quickly and by the early 1980's the country was losing nearly 4% of its forests every year (Carriere, 1990:150).

Costa Rica's links with the capitalist world-economy largely explain the increased deforestation of the 1970's. Like Brazil, Costa Rica has been a peripheral country in the world-economy, a close satellite of the United States due to its proximity. It has been a producer of agricultural commodities and cattle products to be sold in the international markets, especially the U.S. market which by the 1980's absorbed 70% of its exports (U.S. Department of Commerce, 1989:11). Coffee, bananas, and beef have traditionally accounted for half of the country's exports (Edelman, 1983:175). Thus, deforestation in the country can be viewed as a byproduct of the expansion of capitalist agriculture adjusting to global demands, in this case largely U.S. demand. This expansion "includes the reorganization of local, regional, and international systems of production, and consumption in order to facilitate capital accumulation" (Stonich, 1989:273).[3] In the 1970's cattle was the most environmentally damaging activity. While until 1970 most of Costa Rica remained under forest cover, some 26,000 km^2, in the 1970's the rates of deforestation rose substantially due mainly to pasture expansion. In the early 1980's the country's deforestation rate of 4% was the highest in the Western hemisphere. From 1970 to 1980 over 7,000 km^2 of forests were cleared, reducing the forests cover from 51% to 36% of the land mass (Carriere, 1990:150-151). U.S. demand for Costa Rican products to a large extent explains the increased deforestation. The bulk of what was produced by Costa Rica was exported to the United States (Rosene, 1990: 372). Demand alone, however, does not fully explain the quick pace of deforestation.

Costa Rica's policies towards its forests have, until the 1980's, been similar to those of Brazil. Public forest lands were defined as undeveloped lands, available for economic expansion, partly to avert social discontent caused by inequalities of land distribution. As in Brazil, individuals could claim these lands by occupying and "improving" them. The Ley de Cabezas de Familia of 1892 granted title of state wilderness lands to those who cultivated them. This law was basically extended through the Ley de Poseedores en Precario of 1942. The principle of "improving" the land by clearing the forest was maintained (Harrison, 1991:84). Large landowners as well as landless peasants could take advantage of the law. For peasants, frontier land has often been the only viable alternative in a country where there is such disparate land distribution, 61% of the land under cultivation concentrated in 6% of the largest holdings (Carriere, 1990:153).

The purpose of clearing forest lands in Costa Rica has varied with the types of commodities in demand in the international markets. In the 19th century, for example, coffee was critical in this process. In the early 20th century banana plantations played a substantial role. More recently the clearing of virgin forest for cattle pasture has dominated. Cattle ranching rose from 24% of the Frontier Region's total farmland in 1950 to 49% in 1984. Note, however, that beef production has increased both as a result of international and internal demand. Costa Rica's internal consumption grew more than beef exports in the period from 1955 to 1984, from 100,000 heads per year to 150,000 heads (Harrison, 1991:90). Still, the bulk of production is for the international market.

Attention to ecological problems has emerged with unprecedented fervor in the last decade and a half in Costa Rica. The outcry for preservation was first mainly from environmentalists, reporters, and scientists in North America and Europe. In Costa Rica the public and government leaders responded by paying increasing attention to the environment (Thrupp, 1990:245). Democracy has allowed environmentalism and environmental protection to flourish. While Brazil emerged out of the military dictatorship with an authoritarian, "hard-line" view of the environment, guided by the notion of sovereignty espoused by the military, Costa Rica was open to ecological debates. While Brazil saw international criticisms as a form of environmental imperialism, Costa Rica was receptive to assistance from abroad, it did not perceive it as a possible foreign takeover of the country. This difference is well-illustrated on the issue of debt-for-nature swaps. While Brazil refused to participate in such programs, Costa Rica engaged in all sorts of swaps to reduce its US.$4.5 (in 1989) billion dollar debt, engaging in negotiations with organizations such as World Wildlife Fund, The Nature Conservancy, and Conservation International (Thrupp, 1990: 246, 250). The relationship it maintained with the United States helped ease xenophobic fears. It has kept exceptional long-standing ties with the United States government (with one of the highest levels of U.S. aid per capita in the third world). It also has been unusually open to U.S. scientists. The country has been a popular destination for biologists and ecologists who have extracted research data there for many years. Costa Rica has a reputation among northern donors and conservationists as being environmentally conscious and successful (see Thrupp, 1990:245). This relationship has compelled one critic to argue that environmental assistance to Costa Rica has been "politically motivated." While Nicaragua, which contains the largest expanse of rainforest remaining in Central America, was ignored during the Sandinista years by the United States, Costa Rica with "islands of tropical rainforests" received all the attention. For this critic, "communist rainforests apparently do not need to be saved . . . " (Vandermeer, 1991:45). The fact is, however, that Costa Rica was also willing to receive the help and implement changes. While Brazil was not, due to its nationalism. While at the same time, Brazil was selling its natural resources at low prices to service its foreign debt.

Costa Rica more readily adapted itself to changes in global ecopolitics than Brazil. In terms of legislation it took quicker steps in the direction of addressing environmental issues. In 1988 it created the Ministerio de Recursos Naturales, Energias y Minas (The Ministry of Natural Resources and Energy) to consolidate its preservation efforts. This branch of government devised the *Estrategia de Conservacion para Desarollo Sostenible de Costa Rica (MIRENEM)* (Conservation Strategy for the

Development of Costa Rica's Sustainable Development) in 1990. This document formally incorporates principles of sustainable development as part of the development policies of the government (Costa Rica, 1997: Panorama Geral). These and other efforts have paid off. Costa Rica's environmental record has improved enormously since the 1980's, making the country a model of preservation in Latin America. Overall its rates of deforestation declined from about 4% in the 1970's and early 1980's to 2.7% in 1985 and 0.36% in 1994.[4] While 420 Km^2 per year of forests were being cleared by 1985, by 1990 this amount had been cut by nearly half, to 220 km^2 per year. By 1994 it dropped even further to 80 km^2 per year (Costa Rica, 1997). It also managed to increase reforestation from 31 km^2 per year by 1985 to 139 km^2 per year by in 1990 and 147 km^2 by 1994, augmenting its forested areas from 15,071 km^2 in 1985 to 21,657 km^2 in 1994. Of this area 10,945 km^2 (51%) has been reported as protected by the government. This figure represents 21.4% of the national territory (Costa Rica, 1997: Ch.11). It is important to note that another source cited "some 25%" of the country's total land area as "protected" (Jones, 1992:679). Another lists the area under "nature reserves" at 1.5 million acres (607,050 hectares), roughly 11% of the national territory (Burnett, 1998). Our assumption is that the latter data are restricted to national parks alone.

Like Brazil, Costa Rica has also taken steps throughout the years to protect Indian lands with mixed results. By the late 1980's, there was a total of 21 reservations in the country occupying an area of 320,888 hectares and with a population of 25,000 Indians, 1% of the total population (Anonymous, 1988:11). Most reservations were created through Executive Decrees after the promulgation of the Ley Indigena of 1977. According to this law, ". . .the Indian reservations [were] the exclusive property of the aboriginal communities that inhabit them. . ."[5] and were to be governed by the Indian themselves. The law also established that non-Indians could not claim property of these lands unless they had lived there for a long period of time (Anonymous, 1988:39-40). However, reservations continued to be invaded by non-Indians despite the law. A study by Robert M. Carmack in the town of Buenos Aires in 1986 showed that 60% of the inhabitants of Indian reserve lands were non-Indians and that they held over 62% of reservation lands (Carmack, 1989:33). In 1984, the National Commission on Indian Affairs (CONAI) introduced legislation proposing a redefinition of Indian reserves in an attempt to remove non-Indians from them. By the time of Carmack's study in 1986 a town meeting was held in Buenos Aires by a special commission from the National Assembly studying possible legislative revisions of the laws regulating Indian affairs (Carmack, 1989:30).

It is important to note that the Indians in Costa Rica have been fully incorporated into the capitalist world-economy, in Hall's (1986) terminology presented in Chapter 1 they belong to "full blown or dependent" peripheries. This is to a large extent evidenced by their participation in the Costa Rican economy as (migratory) agricultural workers and their degree of cultural assimilation (see Anonymous, 1988). As a result, they are at the bottom of an ethnic pyramid as the most disenfranchised ethnic category (Carmack, 1989:31). They are subject to prejudice and discrimination and viewed by even educated people as "inferior beings" compared to non-Indians (Anonymous, 1988:12). An incident witnessed by anthropologist Philippe Bourgois when studying the Guaymi Indian plantation workers is illustrative of this prejudice:

> Perhaps the best example of the level of public humiliation the Guaymi repeatedly suffer is the scene I witnessed within half an hour of my first arrival at the bus terminal on the Costa Rican side of the plantation--an ape-like howling filled the air "whoo whoo whoo whoo." Young Latin men and women who owned the shanty shack stores surrounding the bus terminal were jeering at an open company transport cart carrying young Guaymi returning from a day's work spreading potassium fertilizer on the Costa Rican farms. The cart stopped and a group of shanty shack owners huddled around the Amerindians, jeering and taunting them. Showing no sign of emotion, the stone-faced Amerindian workers approached the howling shanty sellers and bought soft drinks and candy. After the workers left, the merchants bragged over how much they have short changed the **"cholitos."** Several other Guaymi browsing around the surrounding shops overheard them (Bourgois, 1985:25).

Thus, Costa Rica still has much to do to eradicate discrimination against Native peoples and to fully protect Indian rights over their ancestral lands against the intrusion of non-Indians.

The Indian issue is not the only sore spot in Costa Rica's record. Critics argue that the system of protecting forests through the creation of national parks is failing. These parks suffer from lack of funding which in turn severely limits the hiring of an adequate surveillance personnel. Illegal logging as well as poaching continue to occur in Costa Rica's main national parks. Poaching in some areas have created the phenomenon of "empty forests," forests without animal life. These problems are present despite the fact that entrance fees to the parks brought in U.S.$4 million and ecotourism in general U.S.$700 million in 1997 (surpassing the revenues generated by the export of bananas and coffee). Much of the money generated by the presence of the parks is channeled to sectors other than environmental protection, e.g., for infrastructure improvement such as roads (Burnett, 1998). Thus, the picture is far from perfect. However, despite these problems Costa Rica presents a very advanced model of preservation in the world.

Malaysia

Despite cultural and geographical differences Brazil and Malaysia have in common a recent history of authoritarianism. But, while Brazil has taken very large steps towards democracy in the last decade, Malaysia has clung to an authoritarian regime. It is interesting to note, though, that Malaysia has been described by one scholar, Gordon P. Means (1996) as "soft authoritarianism," a term that has often been used to describe the Brazilian military years, e.g., referring to Brazil as a *ditamole*, or soft dictatorship. As a matter of fact, overall some aspects of the Brazilian dictatorship resemble many features of the government of Malaysia in the present.

Malaysia is a multi-ethnic society composed of 58% Malay and other indigenous, 26% Chinese, 7% Indian, and 9% other (CIA, 1997). The Malays are politically the dominant group, holding the vast majority of seats in parliament. Malaysian politics have been punctuated by Malay efforts to remain the dominant group in a stable multi-ethnic country. Since independence from Great Britain in 1947 Malaysian politics have been characterized by an effort to form ethnic alliances and thus guarantee political stability in the country. It also has been characterized by the fear of ethnic unrest, a fear grounded on periods of ethnic instability such as that which began on May 13, 1969 (see Means, 1991: Ch. 4-18). The possibility of ethnic strife goes a long way in explaining the restrictions on democratic freedoms that prevail in the country. These conflicts have helped corrode the democratic institutions that Malaysia inherited from the period of British rule. For example, parliament exists largely to legitimize the decisions made by the executive. It is dominated by the ruling coalition National Front (Barisan Nasional), a coalition formed by the leaders of different ethnic organizations, but dominated by the United Malay's National Organization (UMNO) (Eldridge, 1996:304). This situation is very similar to that of Brazil in the 1960's and 1970's when Congress was maintained, but controlled by the government's party Aliança Renovadora Nacional (ARENA). Congressional decisions did not count in the end because the executive had actual control of the state. In addition, opposition and civil disobedience in Malaysia are discouraged by the government, a condition prevalent in the Brazilian dictatorship years also. As a result of the ethnic conflicts of 1969, the Malaysian government devised an ideology of social integration or harmony, the *Rukunegara* or Pillars of the Nation. This ideology consists of the following principles:

1) Belief in God
2) Loyalty to King and Country
3) Upholding the Constitution
4) Rule of Law
5) Good Behavior and Morality

While the idea of upholding the constitution would seem to be conducive to democratic values, Malaysia is a theocracy, Islam is the state religion (even though other religions are tolerated), and the law is constantly being ignored by declaring a situation of emergency, thus allowing government to use "emergency powers," including "preventive detentions" (Means 1996; 1991:8-18). To a large extent the country is governed by the principles of Good Behavior and Morality, which means not criticizing the policies of the government. Antagonizing the government means violation of the Rule of Law principle.

Like Brazil's "Brazil: A Superpower in the Year 2000" ideology of the Brazilian military, Malaysia is guided by a strong developmentalist ideology whose success legitimizes government action:

> Inter-religious and community relations have managed with increasing success, buttressed by rapid economic growth in which most groups have felt some benefits. Since 1990, the "New Economic Policy (NEP) explicitly favouring indigenous Malays in business, education and government has been transformed into Dr. Mahatir's more inclusive "2020 Vision" of an advanced industrialized society, in which all groups are to participate harmoniously. In principle, this could favour a gradual democratic evolution. However, underlying tensions between the new Malaysian (*Banga Malaysia*) as against the older Malay (*Bangsa Melayu*) nation concept, for many years advocated by Dr. Mahathir himself, is likely to be seen by the government as justification for maintaining security powers (Eldridge, 1996: 304-305).

These development policies entail the utilization of natural resources. And in Malaysia, perhaps more than any other developing country, tropical rainforests are viewed as natural resources to be utilized for the development of the country. Timber is Malaysia's second biggest export earner after petroleum, and as of 1992 Malaysia accounted for two-thirds of the world export of tropical logs, mostly to Japan (Wolf, 1996:437). Logging provides Malaysia with quick foreign currency that helps to stabilize its economy. And, politicians have no qualms in terms of utilizing these natural resources for what they term development.

While the developmentalist attitudes and the importance of logging to the Malaysian economy have led to deforestation and a political culture of resistance to environmental protection, the financial rewards received by politicians is perhaps a more powerful force. Logging in Malaysia is done through a system of concessions in which large tracks of "State" lands are awarded to individuals who in turn allow their own concessions to logging companies. These "forest barons" are the greatest beneficiaries in the logging system, receiving huge sums of money by virtue of owning a concession. The system is set up in such a way that the "forest barons" are not so easily detected, thus they do not have to be accountable for their actions. The Malaysian environmental group *Sahabat Alam Malaysia* has the following to say about this arrangement:

> It is not always easy to identify the real owner of these timber concessions. Normally, individuals are given grants/licenses, who then commission the licenses to major timber contractors to work the commissioned area. Sometimes the major contractor will sub-contract it out to different contractors. This practice of sub-contracts, forming a kind of hierarchical structure, makes the original licensees virtually "invisible", they may never actually visit the logging camp yet profit enormously from it. Whoever they may be, it is common knowledge that timber licenses are usually owned by politicians or those related by blood or money to them. The main reason for the development of the timber industry is economic. Yet, only a minority of the population benefits from the development (in Hong, 1987:86).

Thus, nepotism and alliances play a key role in the process. Companies do not have to be responsible in their logging operations because they have powerful connections in government. In a study of 12 timber concessions conducted by the Sarawak Study Group in 1985 it was found that many of the most important shareholders were politicians belonging to the ruling party, senior civil servants or close aids to political leaders and their relatives (Hong, 1987:86).

The dynamics of deforestation in Malaysia become very vivid in the state of Sarawak in the island of Borneo. Sarawak's forests are to Malaysia as Amazônia is to Brazil, the frontier. The state occupies 124,449 km^2 (12,444,900 hectares) and has the largest concentration of Malaysian native peoples. The Dyak peoples comprised by 1980 some 44% (539,834) of the population of the state. By December 1984, timber concessions in the state involved 5,752,996 hectares of forests, of which 1,342,826 hectares had already been exploited. By 1985 Sarawak alone accounted for 39% of Malaysia's total log output (Hong, 1987:2, 85). For the Dyak peoples the wealth contained in their forests meant that they would be subjected to the worse forms of capitalist exploitation.

State legislation made it easy for "forest barons" and logging companies to rob the Dyak peoples of their native lands. The state of Sarawak has been divided into five main land categories: 1) Mixed Zone Land; 2) Native Area Land; 3) Native Customary Land; 4) Reserved Land; and 5) Interior Area Land. As the term implies, Mixed Zone Lands may be held in title by diverse categories of people, including indigenous people themselves. Native Area Lands are lands that may be held by natives under title. Non-native people, however, may also acquire rights over them, e.g., by prospectors. Native Customary Lands are lands in which native customary rights whether communal or not have been established before January 1, 1958. After this date, a community can only have rights over the land under this condition through a permit granted by the government. The 1974 Land Code (Amendment) Ordinance gave the right to the government to extinguish native customary rights; that is, from then on the natives could lose their rights over land at government's will. Reserved Lands are those reserved by the government for various purposes. They include forest reserves, protected forests, national parks, wildlife sanctuaries, nature monuments, etc. The government has the right to declare land owned by the state, including those occupied by natives, Reserved Land. Finally, Interior Area Lands are those that do not fall within any of the previous categories. They are located deep in the interior of the state and under primary forest. They cannot be held under title but native customary rights may be created subject to a permit (Hong, 1987: Ch. 4).

The law has had deep consequences for the native people of Sarawak. Legislation allows the government to take away native customary rights and grant native land as concessions to logging companies. This resulted in the decline in the number of native communal forests, from 303 in 1974 to 56 in 1984 (Khan and Khan 1995/1996, p.1). Native people unexpectedly would be visited by officials from logging companies with government permits in hand allowing them to log ancestral lands, often with bulldozers already in the Natives' backyards. These companies have little respect for the Dyak people's traditions:

A week later in the Apoh river, natives of Long Layun who had lived there for over 50 years were told by the manager of a logging camp that they were not allowed to farm the land 'just above our longhouse because the land belongs to the company.' The manager had told them that 'the government owns the land and gave it to us.' These villagers only discovered that the communal forest reserved was occupied when they saw bulldozers coming. Not only was the land taken away, the village cemetery was also desecrated in the process. According to one native: 'They did not bother to show us any respect. They thought we were just wild animals (Hong, 1987:90).

The lack of notification by the government compelled the headman of the Long Adang to state the following:

> Who but thieves come to a place without asking or even telling the owners first. I call these people who have come to our lands thieves (Along Saga in SOS Sarawak, 1991).

This type of situation was pervasive throughout the 1980's, eventually leading the natives to revolt in 1987. In that year Orang ulu tribes began to erect barricades across logging roads in an attempt to stop deforestation and pollution, e.g., of rivers, in their customary lands (Khan and Khan, 1995/1996, p.1). The reaction of the government was to prosecute and imprison the leaders of the movement. When natives attempted to negotiate with government officials they were ignored or jailed. Their demands were obviously too much of a threat to the economic interests of the Malay elites. Like Amazonian Indians such as the Yanomamis they were sitting on what non-natives defined as wealth.

Efforts to help the Dyak peoples have been limited by Malaysian authoritarian laws regulating free association. A new law in 1981, stiffened the requirements for the creation and functioning of "Malaysian societies." To be established these organizations have to go through a lengthy bureaucratic process of recognition by the government. The law requires that organizations which issue public statements be classified as political rather than civil societies, thus subjected to even greater scrutiny. All societies are also required to submit an annual report to the government, which may or not renew their recognized status. The government has full discretion in declaring a society unlawful on security, public order or morality grounds. The law also requires that any organization deemed political to ensure that all of their members and officers are Malaysian. These organizations were also required by law to obtain the Register's written permission to continue or initiate any foreign affiliation or connection (Singh K.S., 1984:2-7). The stiffening of the legislation had much to do with the government's annoyance with dissenting voices:

> Among the organizations which appeared to be the targets of this legislation were the *dakwah* youth movement, ABIM, the modernist reform and consumer movements, Aliran Kesedaram Negara (National Consciousness Movement), and the Consumer's Association of Penang (CAP), which had taken the lead in raising public attention to environmental and ecological issues (Means, 1991:85).

When the situation in Sarawak made international headlines the Malaysian government turned against environmental organizations and the international press. Its solution was an authoritarian solution. It attempted to limit freedom of the press and freedom of speech:

> For several years, the government had assumed a hostile stance towards the operations of the foreign press in Malaysia. The government proceeded on the assumption that the foreign press should respect the Malaysian 'national interest' as defined by the government and therefore news reports should not be biased against the government and not contain any 'unauthorized' information concerning government activities or policies. When the foreign press refused to accede to these unwritten assumptions about their role, a series of restrictions and harassments were placed on their activities (Means, 1991:196-197).

From May 1984 on the Malaysian news agency, Bernama, was made the sole official domestic distributor for all foreign agencies and the sole wire transmitter of news from reporters in Malaysia to their foreign news desks. When these guidelines were not followed the government responded by prosecuting news corespondents and revoking their work visas (Means, 1991:197).

The situation got especially severe for critics of the government's environmental policies as the situation in Sarawak received notoriety abroad. This violated the non-criticism-of-government rule:

> The 'greens' have, in fact, violated a fundamental rule that earlier governed their activities; they have questioned the basic philosophy of national development in taking on the timber industry and allying themselves with well-financed international movements. Moreover, they implicitly threatened the interests of powerful people with influence in very high places (Brookfield, 1994:285).

When the crackdown of those deemed to be "destabilizing the political balance between the nation's ethnic groups" came in 1987, 10 leaders of environmental groups were taken into custody. Even though their imprisonment was brief, the message that the government would not tolerate criticisms of its environmental policies was clear (Brookfield, 1994:285).

The unchallenged power of the government has meant that Malaysian logging companies have had the upper hand in destroying forests. Their reputation for deforestation and ravaging the environment is so bad that they are feared even in Brazil, as already pointed out in Chapter 1. They are notoriously known for their disregard for environmental law in the places where they operate. They have been responsible for severe deforestation in the Malay Peninsula as well as in Borneo. In Peninsular Malaysia, dryland forests declined from 64% of the total area in 1966 to 54% in 1982, and by the early 1990's it was less than 27%. In Sarawak 2,821,711 hectares of forests had been logged by 1985. Of this figure 43% was cut between 1980 and 1985 alone. By 1984 60%, 56,768 km^2, of Sarawak's remaining forests were licensed out for logging (Hong, 1987: 128-131; Brookfield and Byron, 1990:49-50). Edward B. Barbier et al. (1994:6) estimated that between 1981 and 1990 255,000 hectares (2,5550 km^2) of forest were being cleared per year in Malaysia as a whole, a rate of 1.39%. According to the official data presented by the Malaysian government to the U.N. for the "Rio Plus Five" conference, the overall rate of deforestation for Malaysia was 3,100.1 km^2 for 1990. This represents 1.6% of the forest cover for that same year (Malaysia, 1997, Ch.21). The document also claims that 43% (14.2 million hectares) of its total land as Permanent Forest Estate (PFE) to be managed sustainably. It also reported that a total of 118,800 hectares are "... to serve as permanent nature reserves and natural arboreta, conserving various forests and ecological types in their original condition" (Malaysia, 1997; Ch.21). It should be noted, however, that Malay legislation has allowed for boundaries of protected areas to change upon notification in the government Gazette by the ruler or Governor in consultation with the Minister of Science, Technology, and the Environment:

> Somehow these things slip through without the Minister, whatever the level of "consultation", having much chance to muster his forces of objection if he sees such objection justified. For the most part, there is no assurance that what is a reserve today will remain tomorrow (Rubeli, 1982:101).

In addition some parks remain in name only. The Taman Negara National Park, for example, which falls in the states of Pahang, Trengganu, and

Kelantan has been legally reserved by the respective states for public purposes; that is, for logging (Rubeli, 1982:101). A similar situation happened to the Endau-Rompin National Park in the late 1970's:

> For the past three months, seven days a week, and 24 hours a day, scores of trucks have been emerging from the jungles of Pahang, fully loaded with tons of solid hardwood logs and roaring towards Singapore, stirring up storms of dust and controversy all the way. Deep in the jungles of Endau-Rompin activity is at a feverish pitch. Loggers are felling the trees with a vengeance, to recover in the shortest possible time the millions of ringgit they have put into the operation, and to make many millions of riggit more (*Business Times* of December 11, 1977, in Consumer's Association of Penang, 1978:11).

In absolute terms Malaysia's deforestation rate of 3,100.1 km^2 in 1990 does not compare to Brazil's 11,130 km^2 in the same year. However, in relative terms its devastation has been worse. In 1990 Malaysia's forest cover was 193,300 km^2 while Brazil's was 4.9 million km^2, of which 4 million km^2 was in Amazônia (Brazil, 1997; INPE, 1998, 1999; Malaysia, 1997). Brazil's deforestation rate thus represented 0.23% of the total forest area, with Amazônia at roughly 0.30% (1990/1991)(INPE, 1998, 1999). The Malaysian rate of deforestation was 1.6%. This means that due to their size and the rate of deforestation Malaysian forests are more at risk of disappearing than Brazil's. Size, of course, should not be a justification for Brazil to destroy such huge tracks of forest.[6] The point is that high rates of deforestation have a bigger impact on the forests of small countries. Furthermore, if one takes the perspective that each rainforest is unique, whole ecological systems are disappearing faster in smaller countries, thus having a profound impact on the biodiversity of the planet. As a rule of thumb it appears that smaller forests suffer the most. This was certainly the situation for the 1980's:

> Of the tropical countries, Brazil (3.2mn ha) and Indonesia (1.3 mn ha) incur the highest *extent* of annual forest . . . . However, the *rates* of tropical deforestation are highest in those countries that have high annual losses of tropical forest combined with relatively small forests resources, such as Ivory Coast (6.5 per cent), Nigeria (5 per cent), Costa Rica (4 per cent), Paraguay (4.7 per cent) . . . (Barbier et al., 1994:7).

These differences are irrelevant, however, from the perspective of loss of global forest cover and its impact on problems such as the greenhouse effect. The Brazilian situation presents the worst scenario from that perspective. This is perhaps one of the reasons why Brazil is more of a target of international pressure relative to a country such as Malaysia, where entire ecosystems have been destroyed.

Relative to Malaysia, Brazil has taken farther steps towards forest protection. Most of Malaysia's protected forests are under the category Permanent Forest Estates, 14.2 million hectares (142,000 km^2), which are managed forests for the timber industry. Virgin Jungle Reserves comprise only a total of 111,800 hectares (1,118 km^2), which is a little over 0.5% of the forest area. If one were to accept Permanent Forest Estates as preserved areas as the Malaysian government's report implies, Malaysia's total protected area would come close to 74.5% of its forest cover, a fiction that defies the evidence on deforestation presented above. Brazil, on the other hand, has classified roughly 8% (396,000 km^2) of its forests as protected. If one is also to include Indian reservations, 544,500km^2 or 11%, one has a total of 940,500 km^2 or 19% of the Brazilian territory (Brazil, 1997). Of course like the Costa Rican and Malaysian cases this is not a perfect scenario. National Parks and Indian reservations are far from being fully protected from invasion such as the illegal smuggling of mahogany:

> With exhaustion of mahogany reserves on private lands, the smugglers of the most valued wood in Brazil are taking action. They are now operating in military areas and Indian reserves in the Amazon forest, mainly in the south of Pará and north of Mato Grosso. There is no limit to their action. Data provided by FUNAI suggest that at least 60 Indigenous areas in the country are under constant threat by the loggers (*O Globo*, March 12, 1998).[7]

And, as we have seen in the case of the Kayapó reservation, the demarcation of Indian lands does not necessarily mean that Indian themselves will fully preserve the land. The Kayapó have certainly involved themselves in financial dealings with corporations eager to explore resources such as gold and timber contained on their reservation. Brazil, however, seems to be growing more committed to the cause of preservation. Compared to Malaysia it has responded more promptly, albeit sometimes reluctantly, to international criticism. This is the case perhaps because Brazil does not depend so much on the timber trade for its economic survival as Malaysia.[8] This difference is best exemplified by Malaysia's determined attitude when faced with possible retaliation for environmental degradation. When the Austrian government, for example, introduced legislation regulating the import of tropical timber, Malaysia reacted with international protest.

The Austrian legislation was introduced to parliament in 1992 as a reaction from domestic environmental groups. It called for a tax on all products containing tropical timber. It also called for the label "Made from Tropical Wood" or "Contains Tropical Wood" to be placed on tropical imports. The legislation would also implement a certification system which would distinguish tropical timber harvested sustainably. Malaysia reacted to this by threatening to boycott Austrian imports unless the import tariffs and labeling schemes were revoked. It feared that the new legislation could establish a precedent for other European countries to follow. Malaysia also argued that by banning tropical timber Austria was protecting its own markets for temperate timber. It filed a formal protest with GATT's Committee on Technical Barriers to Trade. Malaysia argued that Austria was unilaterally deciding for all members what constituted sustainably managed forest when there was no consensus on the issue. Fearing that the legislation violated GATT agreements the import tax aspect of the legislation was abandoned. The Austrian government also agreed to reevaluate the import-certification measure. It maintained, however, the labeling law on tropical wood products (Wolf, 1996: 438-439).

In contrast Brazil has responded to international pressure by putting a ban on the trade of mahogany. Even though, as pointed out above, this trade continues illegally, there is a growing awareness in government of the damage that this trade has caused.

Indonesia

The situation in Indonesia relative to Brazil is exemplified by the following report in the Worldwide Web entitled "Killing Kalimantan:"

> International pressure played a role in pushing Brazil into action. But no amount of jawboning seems to have any effect in Indonesia. There, president Suharto and large companies, some of them run by his cronies and relatives, have turned the archipelago's once vast and unspoiled rain forests into money machines--and ecological disaster areas. . . . (Colmey et al., n.d.).

Indonesia is an important case study for comparison because it is second only to Brazil in forest expanse. Its forests are thought to cover some 92 to 109 million hectares (Barber, 1997, Section IV, p.1; Birkelund, n.d., p.5). It was also under the control of the military from 1965, when the "New Order" regime of President Suharto began, to 1999, when free elections were finally held. The underlying force behind the Suharto regime was a very powerful developmentalist ideology:

Having risen to power 32 years ago in a country faced with grinding poverty, Soeharto considers himself Indonesia's "Father of Development," and has aggressively and often ruthlessly promoted economic growth (Runyan, 1998, Part 1, p.2).

As in Malaysia this ideology gives priority to development over human rights or individualism:

> He [Suharto was] a key proponent of the "Asian Values" vein of thought, which holds communal harmony in higher esteem than individual rights. He has repeatedly implied that paying heed to human rights and democratization at any expenses to economic growth is a luxury that developing countries like Indonesia cannot afford. "In Indonesia, we respect and carry out the principles of human rights in accordance with our system and our understanding," Soeharto told reporters in response to U.S. criticism . . . (Runyan, 1998, Part 1, p.2).

Until the riots of 1998, the legitimacy of the regime was guaranteed by force and economic performance--as economic performance declined only force remained. From 1968 to 1993 the Indonesian GDP grew by 6%, inflation averaged less than 10%, and the incidence of absolute poverty fell from 60% to 14% of the total population (Barber, 1997, Section III, p. 2). Behind this formidable economic performance has been the quick exploitation of natural resources such as oil and timber.

Of the four countries in this chapter Indonesia is the most authoritarian. Most aspects of its political structure had been controlled by the military during the Suharto years. This included the federal government in Jakarta and the provincial governments[9] in the outer islands. President Suharto's emphasis was on consensus decision making, with open opposition to decisions believed to be politically incorrect. Thus, decisions tended to be conservative and reflective of the status quo (Dauvergne, 1993:505). The role of the military is a major difference between Indonesia and Malaysia:

> Comparing Indonesia and Malaysia, legislatures remain firmly subordinate to the respective executives. . . . However, Malaysian popular representation is based on parties within the governing National Front coalition rather than functional groups. In contrast with Indonesia, the military is subordinate to civilians in Malaysia. Thus, fewer institutional obstacles exist if a clear political will for democratic reform emerges (Eldridge, 1996:304).

As in the Malaysian case, the Indonesian government has been surrounded by charges of corruption at all levels. The timber industry is no exception to this rule. It has been argued that about 25 timber kings ran the timber industry under the patronage of President Suharto (Poffenberger, 1997:461, 466). The power of these tycoons was such that they appealed directly to the president and were above the law:

> On March 8, 1991 [Prajogo Pangestu, president of the Barito Pacific Group, and who is said to personally control 12 million of Indonesia's tropical rainforest worth $5 to $6 billion dollars] sent president Suharto a memo asking that MOF [Ministry of Forestry] paperwork be expedited. The president told the minister to fulfill all of Pajogo's requests, which were soon granted. . . . Further, in July 1991 when the ministry found extensive timber violations by Barito Pacific and levied a $5 million fine, Barito refused to pay and the Ministry was forced to drop the case . . . (Poffenberger, 1997:467).

Indonesian policies towards the environment are similar to those of Malaysia. Tropical rainforests have little value outside their timber value and as potential areas to be developed. This disdain for the environment is exemplified by the reaction of a government official to forest fire:

> Indonesian central government decision makers view forests as a valuable, yet expendable resource, useful for generating foreign exchange to finance industrialization.. The response of the Forestry Minister to the 1983 East Kalimatan fire exemplifies the view that forests are dispensable. In an interview the minister argued that "much of the area that was burned was conversion forest. So what you have is land clearing for free. The forest fire was the natural way of clearing the land...." (Dauvergne, 1993:507).

This attitude resembles the developmentalist views of the Brazilian government during the dictatorship era.

Indonesia has something else in common with Brazil: it has developed schemes to relocate its population to areas it deems unpopulated. Since 1969 the government has resettled 8 million people from the overpopulated islands of Java and Bali in the Outer Islands, e.g., Borneo, in "trans-migration" programs. With the financial assistance of the World Bank these programs have affected some 1.7 million hectares of state forest land (Barber, 1997, Section III, p.10). As we shall see below, the Indonesian government has admitted that these programs are a major factor behind deforestation. They also have had profound implications for the local native populations which are displaced upon the arrival of migrants:

> These sites contribute to the relative scarcities for local populations. Not

only do such local populations lose access to the forests and lands appropriate for these sites, they must compete for resources surrounding the site with new arrivals. This competition intensifies in the many cases where agricultural failure leads transmigrants to seek resources (swidden lands, nontimber products, and game) in adjacent forest. (Of course the government also argues that transmigration *decreases* scarcity on Java and Bali) (Barber, 1997, Section IV, p.11).

Estimates of deforestation for Indonesia are not the most reliable. According to the government of Indonesia, "Data on Indonesia's forests [are] incomplete--a significant complicating factor in sustainable forest management. . ." (Indonesia, 1997).   An estimate puts the rate of deforestation at 1.32 million hectares (13,150 km^2) a year or 1.21% for the period 1981 through 1990 (Barbier et al., 1994:6). Another estimate based on a study by FAO in the late 1980's put the annual rate at 9,000 km^2 (Dauvergne, 1993:497; World Resources Institute, 1992:417, 591), and yet another, by the government of Indonesia, at 1.3 million hectares (13,000 km^2) or 1.2% (see Barber, 1997, Section IV, p.3).   More recently, the government of Indonesia reported in its "National Implementation of Agenda 21" document a forest cover of 140.4 million hectares and deforestation rate of "approximately" one million hectares, roughly 0.71%. It claimed that 67% of the deforestation was caused by state sponsored programs in transmigration, and estate crops and swamp development (Indonesia, 1997: 24).[10] The report has also the following to say about land degradation:

> In rural settings, forest areas at high altitudes and wetlands in the coastal areas are being converted to agricultural uses, with considerable disruption to ecosystem processes. Soil erosion due to these disturbances will cost the country an estimated $300-400 million US per year, 90% of it in the form of loss of land productivity and the remainder in the form of accumulation of sediments in irrigation systems, reservoirs and coastal areas. In Indonesia, the total area of degraded, denuded and waste land is presently estimated at 30 million ha., two thirds of it in Java. Without serious efforts to address this problem, waste land areas can be expected to increase by 1-2% per year (Indonesia, 1997: 23).

The U.N. report also states that 30.7 million hectares (21.9%) of Indonesia's forest areas are classified as Protected, 18.8 million (13.4%) are Nature Reserves or National Parks, 64.3 million hectares (46.8%) are Production Forests and 26.6 million hectares (18.9%) are "Convertible Forests," designated for non-forest uses such as agriculture, settlement and transmigration (Indonesia, 1997). That is, 34.3% of the total forest cover is considered protected and the remaining 65.7% has been targeted for production or to be cleared for development. Closer scrutiny of the data reveals, however, that only 18.7 million hectares of the total 49.5-million-hectare protected area have been "proposed to be gazetted" as conservation area, of a projected 30 million hectares. This figure represents 13.3% of the total forest cover. Also, 30 million hectares of the 49.5 million remain under the ambiguous category "Forest Protection" (see Indonesia, 1997). It is not that other categories would guarantee protection. Protected lands can become "unprotected" overnight. There are no guarantees:

> Yamdena is a small but ecologically valuable island of 535,000 hectares, situated in eastern Indonesia's Tanimbar Islands. In 1971, the government decreed the whole island a protected forest and research area, and expressively prohibited commercial forestry. But in April 1991, the state granted a concession to log 164,000 of the island's 172,000 hectares of forest by P.T Alam Nusa Segar (ANS). Logging began in January 1992, which raised a storm of protest from the island's people. Islanders claimed that the ANS had not consulted the community and the state gave no prior notice that a concession was to be issued. Islanders brought their protest to the national Parliament in July, but logging continued (Barber, 1997, Section IV, p. 14).

Recognizing the volatility of the situation, the Ministry of Forestry froze timber operations on the island for six months. Logging was resumed in August of 1993 despite recommendations by representatives of the Tanibar Intellectuals' Association (ICTI), a local NGO, that the activity would eventually lead to environmental disaster (Barber, 1997, Section IV, p. 14).

Despite its claims of protection, forest cover in Indonesia has declined dramatically since 1950. In that year, the forest cover was estimated to be 152 million hectares. By 1985 it had declined to 119 million hectares, an average of 914 hectares annually. This loss amounted to 33 million hectares, an area the size of Vietnam. It has been estimated that by 1990 the total forest cover was at 108.6 million hectares. And, the government's Sixth Five-Year Development Plan stated that the total area under natural forest in 1993 was only 92.4 million hectares (Barber, 1997, Section IV, p.3; Barbier et al., 1994: 6; Runyan, 1998, Part 3, p.1). Assuming that the deforestation rate of 1 million hectares provided by the Indonesian government to be correct, deforestation as a percentage of forest cover can range from 0.71% in the best scenario (using 104.4 million hectares as a base line) to as high as 1.08% (using 92.4 million hectares). The scenario is a little better than in Malaysia since the total area is larger. It should be kept in mind, however, that Indonesia is composed of more than 17,000 islands, making it the world's biggest archipelago. It stretches 5,000 kilometers from east to west, the same distance as from Baghdad to London. As deforestation concentrates in certain areas it threatens whole ecosystems (Runyan, 1998, Part I, p.2). Regions such as Sumatra, Kalimatan, Sulawesi, Maluku, and Irian Jaya lost 12.6% (2,940,000 ha), 12.3% (4,890,000 ha), 8.3% (940,000 ha), 5.0% (320,000 ha), and 3.7% (1,310,00 ha) of their forests respectively between 1982 and 1990. Sumatra, for example, has lost 70% of its original lowland forests (Barber, 1997, Section IV, p.8).

Resistance to the policies of the government by NGO's and native peoples have been met with hostility:

> The government has pegged the environmental group WALHI (The Indonesian Forum for the Environment)-a coalition of 335 organizations from around the country-as one of 32 "problematic" activist organizations deemed to be "carrying out activities that exceed their charter," writes John McBeth in the *Far Eastern Economic Review*. The military alleges that in 1996, WALHI helped spark riots at the gigantic Freeport gold and copper mine in West Papua, which has close ties to Soeharto. The accusers have offered no evidence to support their claim, and apparently feel no compelling pressure to do so. In Indonesia, where insulting the president can be a capital offense, critics of the country's style of development are forced to tread lightly (Runyan, 1998, Part I, p.4).[11]

Thus, NGO's in Indonesia face problems similar to their counterparts in Malaysia. They are far from being allowed to pose the challenge to the status quo of their Brazilian counterparts. Native resistance to a large extent resembles the situation in Malaysia as well. Outside companies move in with concession papers and force natives out. When they resist, they face imprisonment and or brutality (see Barber, 1997, Section IV, pp. 13-21 for specific examples).

The Overall Scenario

Despite the imprecisions of the data presented above a few generalizations can be made about the four countries. It is evident that both Costa Rica and Brazil have taken the strongest steps in the direction of protecting their tropical forests. Costa Rica has the strongest Preservation Effort of all four countries. From having one of the highest deforestation rates in the world in the 1980's it became a model of preservation in the 1990's, reducing its rates of deforestation from 2.7% a years to 0.36% in 1994. By 1994 it also managed to protect 51% of its forest area, with 11% of the national territory in national parks (Costa Rica, 1997:Ch.11; Burnett, 1998). Among the governments of all four countries its government has been the most receptive to changes towards preservation. To a large, extent, however, this has much to do with the availability of funds coming from countries such as the United States and also through ecotourism. As international interest in tropical rainforests grew beginning in the 1980's, Costa Rica was not only pressured toward preservation it could also count on the financial and technical assistance of the United States. It was, however, receptive to such assistance.

Following in the footsteps of Costa Rica is Brazil. Despite large areas of Amazônia still being cleared it has managed to substantially reduce its rates of deforestation, even though they often oscillate. In the 1990's these rates reached the lowest of 0.30% (1990/1991) for Amazônia and the highest of 0.81% (1994/1995). However, when the rates get too high there is pressure from within Brazil and abroad for the government to take action. In 1995/1996 the rates had already dropped to 0.51%. The country has also managed to protect in one form or another 19% of its national territory. Also, through the New Constitution of 1988 Brazil has made the environment a major issue for its future. The emergence of grassroots groups with democracy in the 1980's means that the government will have a difficult time ignoring constitutional mandates related to the environment. Brazil is far from being an environmental hero, however. By August of 1996, 517,069 Km^2 of Amazônia had been cleared. The measures taken by the government failed to totally stop the destruction. Without increased government support for personnel and equipment for Amazônia, environmental degradation will continue to occur.

In the two other cases, Malaysia and Indonesia have destroyed their tropical forests at accelerated rates and have done the least to avert the destruction. The lack of democratic freedoms in these countries has meant that opposition to the powerful timber lobby is weak. Timber corporations use their ties to government to manipulate the situation through legal and illegal means to their advantage. Environmental organizations are perceived as collaborating with foreign forces against national interests. They are kept under close scrutiny. Indigenous peoples are also not free to demonstrate against the system. In both countries they are thrown in jail for attempting to protect their ancestral lands.

Thus, there is a connection between democratic freedoms and the Preservation Effort of the countries discussed above. As a rule of thumb, democracy allows for all sorts of groups to emerge and challenge the status quo, thus changing the dialectics of power within a given country. It is no coincidence that the dialectics over the environment have changed the most in Costa Rica and Brazil, the most democratic; and the least in Indonesia and Malaysia, the most authoritarian. This is also indicative that changes in global ecopolitics towards sustainable development have had the greatest impact in changing domestic ecopolitics in democratic countries, which are open to debate about new ideas. That is, the type of political system a country has can accelerate or retard the onset of environmentally progressive change.

Despite the importance of democracy, it is important to keep in mind that Brazil and Costa Rica have also been under a great deal more pressure from international organizations to change. It is ironic that while international organizations have put such environmental pressures on Brazil, the same has not been the case for Asian countries.

Notes

[1.] For a discussion on political versus economic democracy or equality see Sidney Verba et al. (1987).

[2.] Our use is an adaptation of Carl Dahlman and Larry Westphal's (1982:105) concept of the "technological effort." These authors define technological effort as being "the use of technological knowledge together with other resources to assimilate or adapt existing technology and/or create new technology."

[3.] It should be kept in mind that this statement was made in conjunction with the situation of Honduras. It is, however, very applicable to the situation of Central America in general.

[4.] Figures for 1985 and 1994 based on data provided by the Government of Costa Rica (Costa Rica, 1997).

[5.] Author's translation.

[6.] It should be noted that this rationale has been used to justify development in Amazônia. It has been argued that the forest is so vast that human beings cannot do enough damage to make it disappear.

[7.] Author's translation.

[8.] Timber is Malaysia's biggest export earner after petroleum (Wolf, 1996:437).

[9.] In 1987, twenty-one of the country's twenty seven governors had ties to the military, they were either retired generals or colonels (Dauvergne, 1993:504).

[10.] It should be kept in mind that official explanations of deforestation are not without their critics. This is the case because "Many tropical country governments...including Indonesia's, blame indigenous people and their traditional farming techniques for deforestation. Although officials claim that some deforestation occur from logging, development projects and "natural" fires, they maintain that the evidence (World Bank data included) clearly indicates that the primary cause is swidden agriculture, commonly labeled as slash-and-burn farming" (Dauvergne, 1993:499)

[11.] This relationship with the government is not new. Trevor Wickham (1987:4) argues that "NGOs in Indonesia face not only an assortment of social and environmental issues but also government restrictions that seek to control the NGOs' autonomy and often threaten their entire existence. To merely survive as an organization requires a considerable amount of finesse and, at times, restraint and compromise."

Chapter 8

Conclusions and Look Ahead

The message of this book is that environmental destruction in the developing regions of the world must be understood in relation to world-system events-- that a national focus is now insufficient. The problem of environmental destruction in Amazônia is that of incorporation into a capitalist world-economy via the Brazilian economy. Amazônia's problems occur wherever this capitalist expansion occurs, as in Malaysia and Indonesia. The forest is being destroyed in the name of progress and civilization. It is viewed only as a source of raw materials; when standing, the forest is viewed as an obstacle. This is culture of development in the modern world-system.

The Brazilian case certainly shows that links with an evolving capitalist world-economy fueled the destruction of Amazônia. Loans from international institutions such as the World Bank and directly from foreign governments to projects such as Carajás sped the implementation of huge projects. It was no coincidence that size increased with each successive project, as international capital became available to Brazil.

No one cared about the fate of the forest and its inhabitants until the hole in the ozone layer and the greenhouse effect shocked the world. Until the 1980's the Amazonian projects were viewed positively, as conquering the world's last frontier. The dominant culture was that regions such as Amazônia had to become economically productive in order to fuel the economic development of the countries which owned them. It is ironic that some of the governments that put pressure on Brazil in the 1990's also financed Carajás. Leaders of the developed countries looked the other way when they believed that their countries would benefit from Brazilian development. There was no accusation by first-world leaders against the military's human rights abuses in the 1960's and 1970's, when Indians were dying in significant numbers, and political dissidents of all sorts were being jailed and killed. Profits prevailed over decency.

Zero-Sum Game and Imperialism

Whereas capitalism operates as if resources have no limits, the earth itself is an enclosed system with limited resources. There is only so much that the earth can absorb in terms of pollution. The first two industrial revolutions, that of Europe and the United States, caused immense global environmental damage. The price of the high standard of living they brought to the West is the global environmental crisis of our time. The world has embarked on a third industrial revolution, that of the Third World. The fact that the earth cannot absorb the impact of another revolution is at the heart of the dispute over regions like Amazônia. Environmental constraints dictate that, with current technology, in order for third-world countries to industrialize, first-world countries will have to substantially reduce their consumption of raw materials and emissions of waste. The greenhouse effect and the hole in the ozone layer simply heightened our awareness of the fact that development as practiced so far cannot continue long without dire consequences. Increased awareness of these implications challenges the dominant views of development. The rhetoric of development has changed.

For developing countries such as Brazil, the environmental pressure from the developed countries seemed hypocritical environmental imperialism. They feared the zero-sum game imposed by environmental constraints.

Change and Dependency

Changes in global ecopolitics towards sustainable development will not come easily. In a world-dominated by capitalists the tendency is to maintain the status quo; changes that may disrupt the system are not welcomed. This becomes clear at every U.N. conference which deals with raw materials and pollution. Corporations lobby fiercely to stop change. This resistance makes it difficult to implement environmental agreements such as those signed at UNCED.

For a country dependent on international investments, like Brazil, changes in global ecopolitics meant that it would have to protect Amazônia to rid itself of its environmental-villain image. These changes were not welcomed. But in the context of the 1980's, when the Brazilian economy was under threat of hyperinflation, and the government was trying to renegotiate its foreign debt, resistance was disadvantageous. International ecopolitics reached its apex with UNCED, and Brazil was the biggest prey. But, Brazil's vulnerability in the world-economy does not fully account for changes in its ecopolitics.

Democracy

When one compares Brazil to Costa Rica, Malaysia, and Indonesia, it is clear that more than the rhetoric of global ecopolitics was necessary to produce internal changes. Countries differ. Malaysia continued to obliterate its forests while Costa Rica chose preservation. While the unique history of each country has to be taken into account, the evidence suggests that democracy is a key component in environmental preservation in the third-world. Democracy exposed Brazilians to an environmental debate in which different sides of the argument could be heard– yes, the US destroyed its forests, killed its native peoples, and continues to pollute the environment, but Brazil does not have to follow that example. With democracy, the cries for human rights from native people could be heard. Brazil wrote a new constitution! Democracy allowed the politics of sustainable development to enter Brazil and Costa Rica, but Malaysia and Indonesia continue to be closed to the idea. While Brazil and Costa Rica have thriving environmentalist movements, environmental groups in Malaysia and Indonesia are thwarted by repression.

As the Brazilian case indicates, the pressure from NGO's in global ecopolitics is critical. Their demands on governments and lending organizations can alter the dynamics of the global system. They have changed the dialectics of power over the environment, internationally and within Brazil.

The political participation of NGO's has also brought some balance to the politics of development, helping democratize the process. International development organizations have been elitist, disregarding the opinions and fate of those mostly affected by development projects. By representing the interests of the disenfranchised NGO's forced these organization to pay attention to the predicament of others. This is important because sustainable solutions involve listening to people who have coexisted with tropical rainforests for thousands of years. Disregarding these voices is to ignore ways of coexisting with nature tested through the ages.

On international and national levels democracy allows those in favor of preservation to be heard. It is by no means a perfect system: those with economic resources influence decisions disproportionately. However, compared to authoritarian regimes, democracy is more conducive to preservation. For those concerned with preserving the environment, the strengthening of democracy is a priority.

Challenges Ahead

The biggest challenge facing capitalism in the 21st is not the threat of communism but the environment. The more the global environment declines the more important it will be to preserve tropical rainforests. The first-world will not convince developing countries to preserve ecosystems if they maintain their current high consumption. Yet, scientific evidence shows that these ecosystems are important for the climatic stability of the planet, that their destruction is contributing to climatic change in ways yet not fully understood. The chaos that climatic change can create has the potential of bringing capitalism to heel. Will capitalism cope with the financial instabilities created by stronger environmental disasters?

It is ironic that in so a difficult an environmental age, we are further globalizing under neo-liberal principles of deregulation. Capital is freer than ever to move transnationally, allowing it to avoid environmental laws. For corporations, the market is the world. What was once consumed only locally now has the potential of being consumed by people throughout the globe, if they can pay the price.

Brazil has embarked on the globalization band wagon; President Fernando Collor saw to that in the early 1990's. One of the consequences is that in times of economic hardship the environment suffers. Brazil encountered rough economic waters in 1998, and nearly drowned in 1999. In order to secure funds from the IMF it had to make drastic cuts in government spending, including reduction in the financing of environmental programs (90% in some cases). And, among the international agencies the IMF is one of the least concerned with the environment. Its billion-dollar package to rescue the Brazilian economy includes no environment clauses. Nor its packages for Asian countries.

So, the challenge for Brazil in the 21st century is to reconcile its developmental needs with the need for environmental conservation. With a growing population and one of the highest levels of inequality in the world, there will be a lot of pressure to further exploit Amazônia. This will put Brazil in direct conflict with international public opinion, especially as this public, faced with growing natural disasters, becomes more aware of the need to preserve the world's remaining forests. Unless the country addresses the tremendous poverty that the bulk of its population lives under, protecting Amazônia will become an insurmountable task. Poor people will continue to flock into Amazônia and put its future in danger if they have no other alternative. Land reform is thus of paramount importance to the survival of Amazônia. Brazil is not a country that lacks land; it is a country where the few have too much land. The wise use of resources includes the question of allocation.

For northeast Brazil, where poverty is prevalent, a preserved Amazônia may be its best hope. The Northeast is a dry region that could become fertile with 15% of the world's freshwater that flows through Amazonian rivers. A system of aqueducts could connect the Amazon to drought-prone states such as Ceará and Piauí– a similar system of irrigation has made California one of the most productive areas in the world. With land and water available to them people will stay in places that are familiar.

Finally, will a vision of the world embodied in sustainable development be enough to tame capitalism? Is the creation of protected areas enough to preserve the ecological functions of tropical rainforests?

Bibliography

Albert. Bruce. 1992. "Indian Lands, Environmental Policy and Military Geopolitics in the Development of the Brazilian Amazon: The Case of the Yanomami. *Development and Change*, 23, 1, 35-70.

Allegretti, Mary Helena. 1994. "Introdução," pp. 17-47, in Ricardo Arnt (ed.) *O Destino da Floresta: Reservas Extrativistas e Desenvolvimento Sustentável na Amazônia*. Rio de Janeiro: Instituto de Estudos Amazônicos e Ambientais, Fundação Konrad Adenauer.

Allen, Elizabeth. 1992. "Calha Norte: Military Development in Brazilian Amazonia." *Development and Change*, 23, 1, 71-99.

Anonymous. 1988. *Reservas Indigenas de Costa Rica*. San Jose:CONAI.

Arruda, Marcos. 1979. "Daniel Ludwig e a Exploração da Amazônia." *Encontros com A Civilização Brasileira*, 11, May, 21-35.

Assis Menezes, Mario. 1994. "As Reservas Extrativistas como Alternativa ao Desmatamento na Amazônia," pp.47-72, in Ricardo Arnt (ed.) *O Destino da Floresta: Reservas Extrativistas e Desenvolvimento Sustentátavel na Amazônia*. Rio de Janeiro: Instituto de Estudos Amazônicos e Ambientais, Fundação Konrad Adenauer.

Bakx, Keith. 1988. "From Proletarian to Peasant: Rural Transformation in the State of Acre, 1870-1986." *The Journal of Development Studies*, 24, 141-160.

Barber, Charles Victor, 1997. "The Case Study of Indonesia," World resources Institute. <Http://www.library.utoronto.ca/www/pcs/state/indo/indonsum.htm>.

Barbier, Edward B. et al. 1994. *The Economics of the Tropical Timber Trade*. London: Earthscan Publications Ltd.

Barbosa, Luiz C. 1993a. "The World-System and the destruction of the Brazilian Rain Forest. *Review*, XVI, 2, 215-240.

_____. 1993b. "The 'Greening' of the Ecopolitics of the World-System: Amazonia and Changes in the Ecopolitics of Brazil." *Journal of Political and Military Sociology*, 21, 107-134.

_____. 1990. "Dependencia, Environmental Imperialism and Human Survival: A Critical essay on the Global Environmental Crisis," *Humanity and Society*, XIV, 4, 329-344.

_____. 1989. "Cotton in 19th Century Brazil: Dependency and Development," unpublished Ph.D. dissertation, University of Washington.

Barbosa, Luiz C. and Hall, Thomas. 1986. "Brazilian Slavery and the World-economy: An Examination of Linkages Within the World-System," *Western Sociological Review*, XVI, 1, 99-119.

Belchimol. Samuel. 1985. "Population in the Brazilian Amazon," pp.39-50, in John Hemming (ed.) *Change in the Amazon Basin, Vol. II: The Frontier After a Decade of Colonisation*. Manchester: Manchester University Press.

Birkelund, James M. n.d. "Chapter Two, Trading Away Forests: Indonesia and Malaysia." <Http://www-personal.umich.edu/~wddrake/545/birkelund.htm>.

Black, Jan Knippers. 1977. *United State Penetration of Brazil*. Philadelphia: University of Pennsylvania Press.

Borges, Beto 1996. "Brazil-Decree#1775 Takes Its Toll," Amazon Program Update, June. <Http://www.ran.org/ran/ran_campaigns/amazonia/ update_6-96.html>.

Bourgois, Phelippe. 1985. *Ethnic Diversity on a Corporate Plantation: Guaymi Labor on a United Fruit Brands Subsidiary in Costa Rica*. Cambridge: Cultural Survival, Inc.

Branford, Sue and Glock, Oriel. 1985. *The Last Frontier: Fighting Over Land in the Amazon*. London: Zed Books.

Brazil. 1997. "Implementation of Agenda 21: Review of Progress Made Since the United Nations Conference on Environment and Development, 1992." United Nations Commission on Sustainable Development, "Country Profiles," <Http://www.un.org/esa/earthsummit/>.

_____. 1988. *Nova Constituição do Brasil*. Rio de Janeiro: Gráfica Auriverde, Ltda.

Brazilian Embassy. 1993. "Brazilian Policy on Indigenous Population," *Current Issues 3*, August.

Brookfield, Harold. 1994. "Chapter 14, Change and the Environment," pp.268-287, in Harold Brookfield (ed.) *Transformation With Industrialization in Peninsular Malaysia*. Kuala Lumpur: Oxford University Press.

Brookfield, Harold and Byron, Yvonne. 1990. "Deforestation and Timber Extraction in Borneo and the Malay Peninsula: The Record Since 1965. *Global Environmental Change*, 1, 1, 42-56.

Buescu, Mircea. 1970. *História Econômica do Brasil*. Rio de Janeiro:APEC.

Bunker, Stephen. 1985. *Underdeveloping the Amazon: Extraction, Unequal Exchange, and the Failure of the Modern State*. Chicago: University of Illinois Press.

_____. 1980. "Forces of Destruction in Amazonia." *Environment*, 22, 7, September, 14-42.

Burnett, John. 1998. "Conservation and Deforestation in Costa Rican Parks," Worldwide Forest/Biodiversity Campaing News, April 3, 1998. Gaia Forest Conservation Archives, Central America. <Http://forests. org/>.

Burns, E. Bradford. 1980. *A History of Brazil*. New York: Columbia University Press.

Cabrera. Any 1996. "Brazil to Review Concession of Indian Lands," in Glen Switkes "Weaken Protection Indian Lands Brazil," <Http://bioc09.uthscsa.edu//natnet/archived/ne/9601/0085.htmt>.

Campuzano, Joaquin Molano. 1979. "As Multinacionais na Amazônia, *Encontros Com a Civilizacão Brasileira*, XI, Maio, 21-34.

Cardoso, Fernado Henrique and Muller, G. 1978. *Amazônia: Expansão do Capitalismo (2ª Edição)*. Editora Brasiliense.

Carmack, Robert. 1989. "Indians in Buenos Aires," *Cultural Survival Quarterly*, 13, 3, 30-33.

Carneiro da Cunha, Manuela. 1987. *Os Direitos do Índio: Ensaios e Documentos*. São Paulo: Editora Brasiliense.

Carriere, Jean. 1990. "The Political Economy of Land Degradation in Costa Rica, *New Political Science*, No.18/19, pp.147-162.

Caufield, Catherine. 1991. *In the Rainforest: Report from a Strange, Beautiful, Imperiled World*. Chicago: The University of Chicago Press.

_____. 1984. "The 62 Billion Question." *New Statesman*, March 23, pp.19-20.

CEDI/Museu Nacional. 1987. *Terras Indígenas no Brasil*. Centro Ecumênico de Documentação e Informação and Museu Nacional, Universidade Federal do Rio de Janeiro.

CEDI/PETI. 1990. *Terras Indígenas no Brasil*, Centro Ecumênico de Documentação e Informação and Projeto Estudos Sobre Terra Indígenas no Brasil. Rio de Janeiro: Museu Nacional, Edição Revista e Atuali dade.

CIA. 1997. "The World Factbook," (under Publications). <Http://www. odci.gov/cia>.

CIMI. 1996a. "World Bank Receives CAPOIB Representatives Lawyers Request Revocation of the Decree." *Newsletter* #196, Feb. 9, 1996. <Http://www.cs.org/clip/cimi/ciminews196.html>.

_____. 1996b. "Resolution of the European Parliament has Repercussions in Brazil," *Newsletter* #198, Feb 23. <Http://www.cs.org/clip/cimi/ ciminews198.html>.

_____. 1996c. "CAPOIB Denounces Brazilian Government; Jobim meets Ambassadors." Newsletter #195. <Http://www.cs.org/clip/cimi/ ciminews195.html>.

_____. 1994a. "Indians Block Roads in Brazil," <Http://bioc09.uthscsa. edu/natnet/archive/nl/9403/0281.html>.

_____. 1994b. "Farmers Threaten to Kill Indians and Missionaries in Brazil," Brazilian Documents Prior to 1997, September, Gaia Forest Conservation Archives <Htt://forests.org/>.

_____. 1994c. "Miners Expelled from Kayapo Land." Brasília, September 22. Brazilian Conservation Documents Prior to 1997, Gaia Conservation Archives <Http://forests.org/>.

_____. 1993a. "There are at Least One-Hundred Indian Organizations in Brazil Already." Brasília, April 15. Brazilian Conservation Documents Prior to 1997, Gaia Forest Conservation Archives <http://forests.org/>.

_____. 1993b. "In 1992, 24 Indians Were Murdered and 24 Others Committed Suicide in Brazil," March 5. Brazilian Conservation Documents Prior to 1997, Gaia Forest Conservation Archives <Http://forests.org/>.

_____. 1993c. "At Least 42 Indians Were Assassinated in Brazil in 1993. Brasília, December 23. Brazilian Conservation Documents Prior to 1997, Gaia Forest Conservation Archives <http://forests.org/>.

_____. 1992a. "Indians in Brazil Keep More than 100 Persons as Hostages to Put an End to Invasion of Their Lands," November 12. Brazilian Conservation Documents Prior to 1977. Brazilian Conservation Documents Prior to 1977. Gaia Forest Conservation Archives <Http://forests.org/>.

_____. 1992b. "Five Thousand Gold Prospectors Have Returned to the Area of the Yanomami Indians." Brasília, November 19. Brazilian Conservation Documents Prior to 1977. Gaia Forest Conservation Archives, <http://forests.org/>.

Citations from the Global Prayer Digest. 1984. "Kayapós." <Http:// calebproject.org/nance/n1723.Htm> (1605 Elizabeth St., Pasadina, CA 94104).

CIVILA. n.d.. "Ciudades Virtuales Latinas, Brasília's Home Page. <Http:// www.civila.com/Brasília/pop_df_i.htm>.

Clark, John. 1991. *Democratizing Development: The Role of Voluntary Organizations*. West Hartford, Connecticut: Kumarian Press.

CNBB. 1986. *Os Povos Indígenas e a Nova República*. Estudos da CNBB 43. São Paulo: Edições Paulinas.

Collins, Randall. 1994. *Four Sociological Traditions*. New York: Oxford University Press.

Colmey, John et al. n.d. "Killing Kalimantan," *Our Precious Planet*. <Http://www.pathfinder.com/time/reports/planet/forests3.html>.

Comitê Interdisciplinar de Estudos sobre Calha Norte, 1988. "Reflexões Sobre o Projeto Calha Norte." *Cadernos do CEAS*, 113, Jan/Fev, 68-77.

Consumers' Association of Penang. 1978. *The Malaysian Environmental Crisis: Selections from Press Cuttings*. Penang: Consumer's Association of Penang.

Correio do Povo, April 10, 1981.

Costa Rica. 1997. "Resena de Costa Rica, Aplicacion Del Program 21: Examen de Los Adelantos Realizados Desde La Conferencia de Las Naciones Unidas Sobrel El Medio Ambiente Y Desarrolo, 1992," Information Presented by the Government of Costa Rica to the United Nations Commission on Sustainable Development. <Http://www.un.org/esa/earthsummit/costr-cp.htm>.

Cultural Survival Quarterly. 1988. "The Calha Norte Project--Multinationals and Mining Militarization in Brazil's Last Frontier," 13, 1, p. 39.

Dahlman, Carl and Westphal, Larry. 1982. "Technological Effort in Industrail Development: An Interpretative Survey of Recent Research," pp. 105-137, in Frances Stewart and Jeffrey James (eds.), *The Economics of New Technology in Developing Countries*. Boulder: Westview Press.

Daly, Herman. 1977. *Steady-State Economics*. San Francisco: W. H. Freeman and Company.

Danaher, Kevin and Shellenberger, Michael (eds.). 1995. *Fighting for the Soul of Brazil*. New York: Monthly Review Press.

da Silva, Aracy Lopes. 1983. "Apresentação Histórica," pp.61-65, Comissão Pró-Índio, *O Índio e a Cidadania*. São Paulo: Brasilience, da Silva, Aracy Lopes et al. 1985. *A Questão da Mineração em Terra Indígena*. Cadernos da Comissão Pró-Índio/Sp No.4. São Paulo.

Dauvergne, Peter. 1993. "The Politics of Deforestation in Indonesia," *Pacific Affairs*, 66, 4, 497-518.

Davis, Shelton. 1993. "Saga of the Yanomami." *Maryknoll,* 87, 4, p. 37.

de Alencar, José. n.d. [1857]. *O Guarani*. Rio de Janeiro: Edições de Ouro.

Dean, Warren. 1983. "Deforestation in Southeastern Brazil," pp.50-67, in Richard Tucker and J. F. Richards (eds.) *Global Deforestation and the Nineteenth-Century World Economy*. Durham, N.C.: Duke Press Policy Studies.

de Arruda, Hélio Palma. 1981. "Razões para a Ocupação da Amazônia." *Estudos de Problemas Brasileiros*, Unidade 51 Textos de Aula, Brasília: Universidade Federal de Brasília.

Degler, Carl N. 1971. *Neither Black Nor White*. New York: Mcmillian Inc.

de Souza Martins, José. 1990. "The political Impasses of Rural Social Movements in Amazonia," pp.245-263, in Davis Goodman and Anthony Hall (eds.), *The Future of Amazonia*. New York: St Martin's Press.

Documentação Indígena Ambiental, n.d. <Http://www.cr-df.rnp.br/~dia/mario.htm>.

Dunlap, Riley, Gullap, George H, and Gullup, Alec M. 1992. *Health of the Planet Survey: Results of a 1992 International Environmental Opinion Survey of Citizens in 24 Nations*. Princeton: The George H. Gallup International Institute.

Dunlap, Riley and Merting, Angela. 1995. "Global Concern for the Environment: Is Affluence a Prerequisite?" *Journal of Social Issues*, 51, 4, 121-137.

Durning, Alan. 1992. *How Much is Enough?* New York: Worldwatch Institute.

Edwards, Michael. 1994. "NGOs in the Age of Information." *IDS Bulletin*, 25, 2, April 1, 117-124.

Eldelman, Marc. 1983. "Recent Literature on Costa Rica's Economic Crisis," *Latin American Research Review*, 18, 2, 166-180.

Eldridge, Philip. 1996. "Human Rights and Democracy in Indonesia and Malaysia: Emerging Contexts and Discourse," *Contemporary Southeast Asia*, 18, 3, 298-319.

Elster, Jon (ed.) 1986. *Karl Marx: A Reader*. New York: Cambridge University Press.

Environmental News Network. 1998. "Brazil Allows Sustainable Logging by Indigenous," February 5. Brazilian Documents After 1997. Gaia Forest Conservation Archives <http://forests.org/>.
_____. 1997. "Brazil Legalizes Indigenous Land Titles," November 29. Brazilian Documents After 1977. Gaia Forest Conservation Archives <http://forests.org/>.
Evans, Peter. 1979. *Dependent Development.* Princeton: Princeton University Press.
Fearnside. Philip. 1990. "Environmental Destruction in the Brazilian Amazon," pp. 179-225, in David Goodman and Anthony Hall (eds.) *The Future of Amazonia: Destruction or Sustainable Development.* New York: St. Martin's Press.
_____. 1989. "Extractive Reserves in Brazilian Amazonia." *BioScience* 39, 387-393.
_____. 1988a. "An Ecological Analysis of Predominant land Uses in the Brazilian Amazon." *The Environmentalist,* 8, 4, 281-300.
_____. 1988b. "Jari ao Dezoito Anos: Lições para os Planos Silviculturais em Carajás." *A Amazônia Brasileira em Foco,* 17, 81-101.
_____. 1986. *Human Carrying Capacity of the Brazilian Rainforest.* New York: Columbia University Press.
_____. 1982. "Jari and Carajas: The Uncertain Future of Large Silvicultural Plantations in the Amazon." *Interciencia,* 7, 6, November-December, 326-328.
Ferreira Reis, Arthur Cézar. 1982. *A Amazônia e a Cobiça Internacional.* Rio de Janeiro: Editora Civilização.
Fisher, William. 1994. "Megadevelopment , Environmentalism, and Resistance: The Institutional Context of Kayapo Indigenous Politics in Central Brazil." *Human Organization,* 53, 3, 220-232.
Folha de São Paulo. 1988. "Organizações Não-Governamentais, Entidades Ajudam Lobby de Esquerda," July 17, 1o. Caderno, A-10.
Fontaine, Pierre-Michael. 1985. *Race, Class and Power in Brazil.* Los Angeles: University of California Press.
Freyre, Gilberto. 1956. *The Masters and the Slaves.* New York: Alfred A. Knopf.
Furley, Peter A. 1990. "The Nature and Sustainability of Brazilian Amazon Soils," pp.309-359, in David Goodman and Anthony Hall (eds) *The Future of Amazonia: Destruction or Sustainable Development.* New York: St. Martin's Press.
Gaia Forest Conservation Archives. <http://forests.org/>.

Gaspari, Elio. 1998. "O Problema Social dos Banqueiros do Jari," *O Globo*, March 15, p.14.

Geisler, Charles and Silberling, Louise. 1992. "Extractive Reserves as Alternatives to Land Reform: Amazonia and Appalachia Compared. *Agriculture and Human Values*, 1 9, 3, 58-70.

Ghazi, Polly. 1994. "The Continuing Fight for the Amazon." *World Press Review*, September, p.46.

Goldemberg, José and Durham, Eunice Ribeiro. 1990. "Amazônia and National Sovereignty." *International Environmental Affairs*, 2, 1, Winter, 22-39.

Goldstein, Karl. 1992. "The Green Movement in Brazil." *Research in Social Movements, Conflicts, and Change*, 2, Supplement 2, 119-193.

Graham, Douglas H. and de Hollanda Filho, Sergio Buarque. 1971. *Migration, Regional and Urban Growth and Development in Brazil: A Selective Analysis of the Historical Record, 1872-1970, Volume I*. São Paulo: Instituto de Pesquisas Econômicas (U.S.P.).

Grubb, Michael et al. 1993. *The 'Earth Summit' Agreements: A Guide and Assessment*. London. Earthscan Publications Ltd.

Guppy, Nicholas. 1984. "Tropical Deforestation: A Global View." *Foreign Affairs*, LXII, 4, Spr., 928-65.

Hall, Anthony L. 1989. *Developing Amazonia: Deforestation and Social Conflict in Brazil's Carajas Programme*. Manchester: Manchester University Press.

_____. 1987. "Agrarian Crisis in Brazilian Amazonia: The Grande Carajas Programme." *The Journal of Development Studies*, XXIII, 4, July, 522-552.

Hall, Thomas. 1986. "Incorporation in the World-System:Toward a Critique." *American Sociological Review*, 51, 390-402.

Harrison, Susan. 1991. "Population Growth, Land Use and Deforestation in Costa Rica, 1950-1984. *Interciencia*, 16, 2, 83-93.

Hecht, Susanna B. 1989. "Murder at the Margins of the World." *Report on the Americas*, XXIII, 1, May, 36-38.

_____. 1985. "Environment, Development and Politics: Capital Accumulation and the Livestock Sector in Eastern Amazonia." *World Development*, 13, 6, June, 663-684.

Hecht, Susanna and Cockburn, Alexander. 1989. *The Fate of the Forest: Developers, Destroyers and Defenders of the Amazon*. London: Verso.

Helmore, Kristen. 1988. "The World Bank and the World's Poor. *Christian Science Monitor*, March 23.

Hemming, John. 1987. *Amazon Frontier: The Defeat of the Brazilian Amazon Indians*. London: Macmillian.

Hohlfeldt, Antônio and Hoffman, Assis. 1982. *O Gravador do Mário Juruna*. Série Depoimentos 2. Pôrto Alegre, Rio Grande do Sul: Mercado Aberto Editora e Propaganda, Ltda.

Holden, Constance. 1988. "The Greening of the World Bank." *Science*, 240, June 17. p.1610

_____. 1987. "World Bank Launches New Environmental Policy," *Science*, 236, May 15, p. 769.

Hong, Evelyne. 1987. *Native os Sarawak: Survival in Borneo's Vanishing Forest*. Malaysia: Institut Masyrakat.

Hopper, David W. 1988. "The Seventh World Conservation Lecture the World Bank's Challenge: Balancing Economic Need With Environmental Protection." *The Environmentalist*, 8, 3, 165-175.

House, Richard. 1989. "Brazil Declines Invitation to Conference on Ecology." *Washington Post*, March 4, A20.

Howe, Marvine. 1975. "Brazil Refuses to Punish Indians." *The New York Times*. January 9.

Hyman, Randall. 1988. "Rise of the Rubber Tappers." *International Wildlife*, 18, 5, 24-28.

Indonesia. 1997. "Implementation of Agenda 21: Review of Progress Made Since the United Nations Conference on Environment and Development, 1992," Information Provide by the Government of Indonesia to the United Nations Commission on Sustainable Development, Fifth Session. <Http://www.un.org/esa/earthsummit/indon-cp.htm>.

Informativo FFCN, 1988. No. 48, October-December.

Ingram, Helen and Mann, Dean E. 1989. "Interest Groups and Environmental Policy," pp. 135-157, in James P. Lester (ed.), *Environmental Politics and Policy*. Durham: Duke University Press.

INPE. 1999. "Monitoring the Brazilian Amazon Forest." <Http://www.inpe.br/>.

_____. 1998. "Divulgação das Estimativas Oficiais do Desflorestamento Bruto na Amazônia Brasileira–1995, 1996 e 1997." <Http://www.inpe.br/amz.htm>. The address for INPE is <Http://www.inpe.br>.

Inter-American Development Bank, United Nations Development Programme, and Amazon Cooperation Treaty. 1992. *Amazonia Without Myths*. (No place of publication nor publisher given).

InterPress Service. 1996. "Brazil: NGOs Protest Cardoso Move on Indian Lands," February 02. <Http://bioc02.uthscsa.edu/natnet/archive/nl/9602/0024.html>.

Irwin, Howard. 1977. "Coming to Terms With the Rainforest," *Garden*, 1, 2, 29-33.

James, C. L. R. 1980. *Notes on Dialectics: Hegel, Marx, Lenin.* London: Allison and Busby Limited.

Jobim, Danton. 1970. "O Problema do Índio e a Acusação de Genocídio," Conselho de Defesa dos Direitos da Pessoa Humana, Ministério da Justiça. *Boletim*, No. 2.

Jones, Jeffrey. 1992. "Environmental Issues and Policies in Costa Rica: Control of Deforestation. *Policy Study Journal*, 20, 4, 679-694.

Jornal do Brasil: March 19, 1998; March 6, 1998; June 13, 1992; March 28, 1991; July 17, 1990; July 15, 1990; July 14, 1990; July 12, 1990; July 8, 1990; July 2, 1990; June 29, 1990

Jornal do Brasil on Line. 1998. "Índios Invadem Prédio da Funai em Campo Grande," April 8. <Http://www.jb.com.br/extra/ex0804939.html>.

_____. 1977. "Justiça Devolve Area de 778 Hectares a Patoxós," May 2. <Http://www.jb.com.br/brasil.html>.

Juruna, Mário. 1986. *Discurso da Liberdade II, 1985-1986.* Mário Juruna Deputado Federal, Câmara dos Deputados. Brasília: Centro de Documentação e Informação, Coordenação de Publicações.

_____. 1984. *Discursos da Liberdade 1983/1984.* Discursos e Projetos de Lei Apresentados pelo Deputado Mário Juruna. Brasília: Coordenação da República.

Keck, Margaret E. 1992. *The Workers' Party and Democratization in Brazil.* New Haven, CT: Yale University Press.

Khan, Mizan and Khosla. 1995/1996. "Dayaks in Malaysia: Major Developments Since 1990," <Http://www.bsos.umd.edu/ cidcm/mar/maldayak.htm>.

Kinkaid, Gwen. 1981. "Trouble in D.K." *Fortune*, April 20.

Koch, Matthias and Grubb, Michael. 1993. "Chapter 9, "Agenda 21," pp.97-157, in Michael Grubb et al. *The 'Earth Summit' Agreements: A Guide and Assessment.* London: Earthscan Publications Ltd.

Kohlhepp, Gerd. 1981. "Ocupação e Valorização da Amazônia: Estratégia de Desenvolvimento do Governo Brasileiro e Empresas Privadas. *Revista Geografica del Institute Panamericano de Geografia e Historia*, 94, July-December, 67-88.

Kolk, Ans. 1996. *Forests in International Environmental Politics: International Organizations, NGOs and the Brazilian Amazon.* Utrecht, The Netherlands: International Books.

Kraft, Michael E. and Vig, Norman J. 1990. "Environmental Policy from the Seventies to the Nineties: Continuity and Change," pp. 3-31, in Norman J. Vig and Michael E. Kraft (eds.), *Environmental policy in the 1980s: Towards a New Agenda.* Washington, D.C.: CQ Press.

Kubitschek, Juscelino. 1975. *Por que Costruí Brasília.* Rio de Janeiro: Edições Bloch.

Landim, Leilah. 1988. *Sem Fins Lucrativos: As Organizações Não-Governamentais no Brasil.* Rio de Janeiro: Instituto de Estudos da Religião (ISER).

Lang, James. 1982. *Portuguese Brazil: The King's Plantation.* New York: Academic Press.

Lemonick, Michael D. 1989. "Feeling the Heat." *Time,* January 2.

Lerner, Steve. 1991. *Earth Summit: Conversations With Architects of an Ecologically Sustainable Future.* Bolinas, California: Common Knowledge Press.

Livernash, Robert. 1992. "The Growing Influence of NGOs in the Developing World. *Environment,* 34, 5, June 01, 12-43.

Lyons, Graham et al. 1995. *Is the End Nigh?* Aldershot, England: Avebury.

Mahar, Dennis J. 1990. "Policies Affecting Land Use in the Brazilian Amazon." *Land Use Policy,* 7, 1, January, 59-69.

_____. 1989. *Government Policy and Deforestation in Brazil's Amazon Region.* Washington, D.C.: The World Bank.

_____. 1982. "Instituições Internacionais de Empréstimo Público e o Desenvolvimento da Amazônia Brasileira: a Experiência do Banco Mundial." *Revista de Administração Pública,* 16, 4, 23-38.

Malaysia. 1997. "Implementation of Agenda 21: Review of Progress Made Since the United nations Conference on Environment and Development, 1992." Information Provided by the Government of Malaysia to the United Nations Commission on Sustainable Development, Fifth Session. <Http://www.org/esa/earthsummit/malay-cp>.

Manchester, Alan K. 1933. *British Preeminence in Brazil.* Chapel Hill: The University of North Carolina Press.

Martine, Georg. 1990. "Rondonia and the Fate of Small Producers," pp. 23-48, in David Goodman and Anthony Hall (eds.) *The Future of*

Amazonia: Destruction or Sustainable Development? New York: St. Martin's Press.

Marx, Karl. 1906. *Capital.* New York: The Modern Library.

Maybury-Lewis, David. 1989. "Indians in Brazil: The Struggle Intensifies. *Cultural Survival Quarterly,* 13, 1, 2-4.

McGrew, Anthony. 1992. "Ch. 1, Conceptualizing Global Politics," pp. 1-28, in Anthony McGrew and Paul Lewis et al. *Globalization and the Nation State.* Cambridge: Polity Press.

Means, Gordon P. 1996. "Soft Authoritarianism in Malaysia and Singapore," *Journal of Democracy,* 7, 4, 103-117.

_____. 1991. *Malaysia Politics: The Second Generation.* Singapore: Oxford University Press.

Mendes, Chico. *Fight for the Forest: In His Words.* London: Latin American Bureau.

Mitchell, Robert C. 1990. "Public Opinion and the Green Lobby: Poised for the 1990s?," pp.84-124, in Norman J. Vig and Michael E. Kraft (eds.), *Environmental Policy in the 1990s.* Washington, D.C.: CQ Press.

Moran, Emilio F. 1981. *Developing the Amazon.* Bloomington: Indiana University Press.

Mougeot, L. J. A. 1985. Alternative Migration Targets and the Brazilian Amazonia's Closing Frontier, pp.50-90, in J. Hemming (ed.) *Change in the Amazon Basin, Vol. II: The Frontier After a Decade of Colonisation.* Manchester: Manchester University Press.

Munn, R. E. 1988. "Strategies for Sustainable Economic Development," *Scientific American,* September, 155-165.

Murphy, Raymond. 1994. *Rationality and Nature: A Sociological Inquiry into a Changing Relationship.* Boulder: Westview Press.

Murrieta, Julio Ruiz and Rueda, Rafael Pinzon. 1995. *Extractive Reserves.* IUCN- World Conservation Union, Commission of the European Communities, Centro Nacional para o Desenveolvimento Sustentado das Populações Tradicionais.

Nando.net. 1996. "Brazil Police Kill 23 Landless in Amazon Class." <http://www.nandotimes.com/newsroom/ntn/world/041896_2307.html>.

Nogueira-Batista, Paulo. 1989. "The Impact of the Environment on the Human Condition: An International Human Rights Problem." Lecture Delivered by His Excellency Ambassador Paulo Nogueira-Batista Permanent Representative of Brazil to the United Nations, on the Social

Work Day at the United Nations. Brazilian Mission to the United Nations.

Novaes da Mota, Clarice. 1992. "Being and Becoming Indian: The Case of the Shoko and Kariri-Shoko of Northeast Brazil. *Proteus*, 9, 1, 26-31.

O'Connor, Geoffrey. 1997. *Amazon Journal: Dispatches from a Vanishing Frontier.* New York: Dutton

O Estado de São Paulo: March 1, 1989; August 26, 1987; August 13, 1987; August 12, 1987; August 11, 1987; August 9, 1987

O Estado de São Paulo.NetEstado. 1997. "Brasília." <Http://www1. estado.com.br/English/Brasil/brasili1.html>.

O Globo: March 16, 1998; March 12, 1998; March 10, 1998, March 8, 1998.

Oliveira, Malu. 1993. "Marajá da Selva." *Isto É*, July 14, No. 1241.

Oliveira Filho, João Pacheco. 1990. "Frontier Security and the New Indigenism: Nature and Origins of the Calha Norte Project," pp. 155-176, in David Goodman and Anthony Hall (eds.) *The Future of Amazonia: Destruction or Sustainable Development?* New York: St. Martin's Press.

Onimode, Bade. 1989. "Introduction," pp.1-8, in Bade Onimode (ed.), *The IMF, the World Bank and the African Debt, Vol. 1, The Economic Impact.* London: Zed Books.

Paiva, Eunice and Junqueira, Carmen. 1985. *O Estado Contra o Índio. Texto 1, Programa de Estudos Pós-Graduados em Ciências Sociais.* São Paulo: Pontifícia Universidade Católica de São Paulo.

Partido Verde (PV). n.d. "The Brazilian Green Party Historical Summary." <Http://www.pv.org.br/summary.html>.

_____. 1986. "Manifesto do Partido Verde." <Http://www.pv.org. brmanif _pv.htm>.

Pereira Gomes, Mercio. 1988. *Os Índios e o Brasil: Ensaios Sobre um Holocausto e Sobre uma Nova Possibilidade de Sobrevivência.* Petropólis: Vozes.

Pinto, Lucio F. 1988. "Calha Norte: The Special Project for the Occupation of the Frontiers." *Cultural Survival Quarterly*, 13, 1, 40-41.

Poffenberger, Mark. 1997. "Rethinking Indonesian Forest Policy," *Asian Survey*, XXXVII, 5, 453:469.

Polak, Fred. 1961. *The Image of the Future: Enlightening the Past, Orienting the Present, Forecasting the Future*, Volume 1. New York: Oceana Publications.

Price, David. 1985. "The World Bank vs Native Peoples: A Consultant View. *The Ecologist*, 15, 1/2, 73-77

Princen, Thomas and Finger, Matthias. 1995. *Environmental NGOs in World Politics: Linking the Local and the Global*. London: Routledge.

ProRegenwald. 1994. "Indians of Roraima Block Road to Stop Miners, March 15". Gaia Forest Conservation Archives, Brazilian Documents Prior to 1997. <Http://forests.org/>.

Rainforest Action Network. 1996. "Yanomami Territory is Again Invaded by Gold Miners." <Http://www.ran.org/ran/ran_campaigns/brazil/Yanomami_update.html>.

Ramos, Alcida Rita. 1997. "The Indigenous Movement in Brazil: A Quarter Century of Ups and Downs." *Active Voices: The Online Journal of Cultural Survival*, <Http://www.cs.org/cs%20website/LeveFour/LevelFour-Ramos>. Originally Appeared in *Cultural Survival Quarterly*, 21, 2, Summer.

Raymond, Nicholas. 1991. "The 'Lost Decade' of Development: The Role of Debt, Trade, and Structural Adjustment," pp. 112-121, in *Global Issues 93/94*. Sluice Dock, Guilford, Connecticut: Dushkin Publishing Group.

Redclift, Michael. 1987. *Sustainable Development: Exploring Contradictions*. London: Routledge.

Robertson, Roland. 1992. *Globalization: Social Theory and Global Culture*. London: Sage Publications.

Roet, Riordam. 1984. *Brazil: Politics in Patrimonial Society*. New York: Praeger.

Rosene, Chris M. 1990. "Modernization and Rural Development in Costa Rica: A Critical Perspective," *Canadian Journal of Development Studies*, X, 2, 367-374

Rubeli, Ken. 1982. "National Parks and Wildlife Reserves: An Attacking Role for Nature's Defenders," pp. 101-105, in Consumers' Association of Penang; the School of Biological Sciences, Universiti Sains Malaysia; and Sahabat Alam Malaysia *Development and the Environmental Crisis: Proceedings of the Symposium "The Malaysian Environmental Crisis."* Penang: The Consumers' Association of Penang.

Ruckelshaus, William D. 1989. "Toward s Sustainable World." *Scientific American*, September, 166-174.

Runyan, Curtis. 1998. "Indonesia's Discontent," *World.Watch Magazine*, May-June 1998. <Http://www.worldwatch.org/mag/1998/98-3e.html>.

San Francisco Chronicle: September 25, 1989; October 1, 1989; April 7, 1989, February 11, 1989; October 13, 1988.

Sanction, Thomas A. 1989. *Time*, January 2.

Schwartzman, Stephen. 1991. "Deforestation and Popular Resistance in Acre: From Local Social Movement to Global Network," *The Centennial Review*, 25, 2, 397-422.

Simons, Marlise. 1987. "Charring the Amazon." *The San Francisco Chronicle*, June 17, 1987.

Singh KS, Gurmit. 1984. *Malaysian Societies: Friendly or Political?* Singapore: Environmental Protection Society Malaysia Selangor Graduates Society.

Skidmore, Thomas. 1974. *Black Into White: Race Relations in Brazilian Thought.* New York: Oxford University Press.

_____. 1967. *Politics in Brazil, 1930-1964: An Experiment in Democracy.* New York: Oxford University Press.

Smith, Nigel J. 1982. *Rainforest Corridors: The Transamazon Colonization Scheme.* Berkeley: University of California Press.

SOS Sarawak. 1991. "Solidarity With Sarawak Penan," *Native-L* (July-August). <Http://bioc09.uthscsa.edu/natnet/archive/n1/91b/0021.html>.

Stern, Paul, Young, Oran R., and Druckman, Daniel. 1992. *Global Environmental Change.* Washington, D.C. National Academy Press.

Stonich, Susan C. 1989. "The Dynamics of Social Processes and Environmental Destruction: A Central American Case Study," *Population and Development Review*, 15, 2, 269-296.

Switkes, Glen. 1996. "Indians Ask Suspension G-7 Funding," February. <Http://bioc02.uthscsa.edu/natnet/archive/nl/9602/0025.html>.

Switzer, Jacqueline Vaughn. 1994. *Environmental Politics: Domestic and Global Dimensions.* New York: St. Martin's Press.

Tarlock, Dan. 1992. "The Role of Non-Governmental Organizations in the Development of International Environmental Law. *Chicago-Kent Law Review*, 68, 1, 61-76.

The Economist. 1989. "Brazil and the Amazon: It is Our Forest to Burn," March 11, 42-43.

The New York Times: January 2, 1994; February 5, 1990; February 25, 1990; November 18, 1981; March 11, 1971; December 8, 1970; June 12, 1970; September 5, 1969.

The Washington Post, March 4, 1989.

Thomas, Caroline. 1992. *The Environment in International Relations.* London: The Royal Institute of International Affairs.

Thompson, Koy. 1993. "Chapter 8, "The Rio Declaration on Environment and Development," pp.85-95, in Michael Grubb et al. *The 'Earth Summit' Agreement: A Guide and Assessment.* London: Earthscan Publications Ltd.

Thrupp, Lori Ann. 1990. "Environmental Initiatives in Costa Rica: A Political Ecology Perspective," *Society and Natural Resources,* 3, 3, 243-256.

Time: September 18, 1989; January 2, 1989

Torres, Alberto. 1914. *O Problema Nacional Brasileiro.* Rio de Janeiro: Imprensa Nacional.

Trade and Environmental Data base. n.d. "Brazil Gold-Mining and Environment (Bragold Case)." <Http://gurukul.ucc.American.edu/ted/bragold.htm>.

Treece, David, 1990. "Indigenous People in Brazilian Amazonia and the Expansion of the Economic Frontier," pp.264-287, in Davis Goodman and Anthony Hall (eds.), *The Future of Amazonia.* New York: St. Martin's Press.

_____. 1989. "The Militarization of Amazonia: The Calha Norte and Grande Carajas Programmes," *The Ecologist,* 19, 6, 225-228.

_____. 1987. *Bound in Misery and Iron.* London; Survival International.

Turner, Terence. 1992. "Defiant Images: The Kayapo Appropriation of Video." *Anthropology Today,* 8, 6, 5-16.

U.N. 1997. "Earth Summit+5." <Http.un.org/esa.earthsummit>.

UNCED. 1992. *The Global Partnership: A Guide to Agenda 21.* Geneva: UNCED.

U.N. Chronicle. 1993. June, 47-66.

Union of International Organizations. 1995. "Changing Relationship between international Non-Governmental Organizations and the United Nations." <Http://www.uia.org/uiadocs/unngos.htm>.

U.S. Department of Commerce. 1989. *Foreign Economic Trends and Their Implications for the United States: Costa Rica.* Prepared by the American Embassy in San Jose, August.

U.S. Government. 1989. "Environmental Impact of the World Bank Lending Volumes I and II." Subcommittee on International Policy and Trade of the Committee on Foreign Affairs, House of Representatives, One Hundred First Congress, First Session, October 4. Washington: U.S. Government Printing Office.

Vandermeer, John. 1991. "Environmental Problems Arising from National Revolutions in the Third World: The Case of Nicaragua," *Social Text*, 28, 39-45.

Veja: September 1, 1993; October 13, 1993; August 25, 1993; April 28, 1993; March 20, 1991.

Velho, Octavio. 1984. "Por Que se Migra na Amazônia." *Ciência Hoje.* 2, 10, 35-39.

Verba, Sidney et al. 1987. *Elites and the Idea of Equality: A Comparison of Japan, Sweden, and the United States.* Cambridge, Massachusetts: Harvard University Press.

Viola, Eduardo. 1988. "The Ecological Movement in Brazil (1974-1986): From Environmentalism to Ecopolitics." *International Journal of Urban and Regional Research*, 12, 2), June, 211-228.

von Puttkamer, Jesco. 1988. "Rondonia's Urueu-Wau-Wau." *National Geographic*, 17, 6, 801-817.

Wallerstein, Immanuel. 1995. *After Liberalism.* New York: The New Press.

_____. 1990. *Unthinking Social Sciences.* Cambridge: Polity Press.

_____. 987. *Historical Capitalism.* New York: Verso.

Wanick, Joby. 1997. "Climate Pact Rescued in Final Hours," *The Washington Post*, December 13, p. A 01

Waters, Malcolm. 1995. *Globalization.* London: Routledge.

Weiss, Zezé and Weiss, Martin D. N.D. "The Yanomami and the Role of a Powerful Anti-Native Alliance. New York: Amanaka'a Amazon Network.

Weston, Roy. 1995/1996. "Toward Better Understanding the Concept of Sustainable Development," <Http://www.rfweston.com/sd/long/welcome.htm>.

Wickham, Trevor. 1986. "Walhi: Indonesia's Environmental Forum," *Environment* 29, 7, 2-4.

Willetts, Peter. 1996. "From Stockholm to Rio and Beyond: The Impact of the Environmental Movements on the United Nations Consultative Arrangements for NGOs." *Review of International Studies*, 22, 1, 57-80.

Wolf, Heather A. 1996. "Deforestation in Cambodia and Malaysia: The Case for an International Legal Solution," *Pacific Rim Law and Policy Journal*, 5, 2, 429-455.

Worcman, Nina Broner. 1990. "Brazil's Thriving Environmental Movement." *Technology Review*, October, 42-51.

World Resources Institutes. 1992. *The Environmental Almanac.* Boston: Houghton Mifflin Company.

Zirker, Daniel. 1991. "The Civil Mediators in Post-1985 Brazil. *Journal of Political and Military Sociology,* 19, 1, 47-73

Zirker, Daniel and Henberg, Marvin. 1994. "Amazonia: Democracy, Ecology, and Brazilian Military Prerogatives in the 1990s. *Armed Forces and Society,* 20, XX, 259-280.

INDEX

Author's Biographical Sketch

Luiz C. Barbosa is an Associate Professor with the Department of Sociology at San Francisco State University. He received his Ph.D. in sociology from the University of Washington in 1989. His interests include environmental sociology, global sociology, social change and development, comparative ethnic relations, and Latin America. He has several journal articles on the Brazilian Amazon rainforest (see Bibliography in this volume).